NEW TECHNOLOGIES IN PROTECTIVE COATINGS

Edited by **Carlos Giudice**
and **Guadalupe Canosa**

New Technologies in Protective Coatings
http://dx.doi.org/10.5772/65573
Edited by Carlos Giudice and Guadalupe Canosa

Contributors

Peter Hammer, Jorge Uruchurtu, Alba Covelo, Carmina Menchaca, Miriam Flores, Pilar Rodríguez-Rojas, Miguel Hernandez-Gallegos, Esteban Martinez-Meza, Rebeca Jaimes-Ramirez, Thirumal Mariappan, Jorge Morales-Hernandez, Hector Dorantes-Rosales, Joel Moreno, Ma. Lourdes Montoya, Aldo Gago, Philipp Lettenmeier, Andreas Friedrich, Carlos Alberto Giudice

Published by InTech
Janeza Trdine 9, 51000 Rijeka, Croatia

Publishing Process Manager Marijana Francetic
Technical Editor SPi Global
Cover InTech Design team

First published August, 2017
Printed in Croatia
Legal deposit, Croatia: National and University Library in Zagreb

Additional hard copies can be obtained from orders@intechopen.com

New Technologies in Protective Coatings, Edited by Carlos Giudice and Guadalupe Canosa
p. cm.
Print ISBN 978-953-51-3491-6
Online ISBN 978-953-51-3492-3

For EU product safety concerns:
IN TECH d.o.o., Prolaz Marije Krucifikse Kozulić 3, 51000 Rijeka, Croatia,
info@intechopen.com or visit our website at intechopen.com.

Meet the editors

Carlos Alberto Giudice, Ph.D., is affiliated with La Plata National University, Argentina. His professional studies involve Chemical Engineering, specializing in paints and coatings as well as anticorrosive and antifouling paints, fire-retardant treatments, heavy-duty offshore protection technologies. He currently serves as Professor of "Materials Protection," director of career doctor in Engineering Materials mention, and secretary of Science and Technology and Postgraduate, National Technological University—La Plata Regional Faculty, Argentina. Previously, he has worked as Principal researcher of CONICET (Scientific Research Career of the National Council for Scientific and Technical Researches, Argentina) Developed work.

He has authored and co-authored more than 250 research articles published in international scientific and technological journals and congress proceedings, about 40 chapters of books and more than 50 advisory works for coating industry, and edited or co-edited 5 books. He served as a lecturer and professor of numerous courses in Latin America, Europe, and Asia.

Guadalupe Canosa, Ph.D., is affiliated with National Technological University, Argentina. Her professional studies include Chemical Engineering, specializing in paints and coatings and technologies such as anticorrosive paints, fire-retardant treatments, and biodeterioration. She currently serves as a Professor of "Physical Chemistry" and "Introduction to Nuclear Energy". She is also a coordinator of career doctor in Engineering Materials mention, executive advisor of the School of Advanced Studies in Engineering Sciences, member of the Science and Technology and Postgraduate Advisory Council. She holds a position of academic counselor of the Chemical Engineering Career (National Technological University—La Plata Regional Faculty), and associate researcher of CONICET (Scientific Research Career of the National Council for Scientific and Technical Researches, Argentina).

She has authored or co-authored several books, chapters of books and research articles published in international scientific and technological journals, congress proceedings and numerous advisory works.

Contents

Preface

The title of this book has been chosen to reflect the combination of basic knowledge and applied science on paints and coatings. The dizzying advance of the science of materials prompted editors to try to produce a book that occupies that space with a scientific and technological approach.

The book is also intended to raise awareness of many underlying knowledge aspects for the successful use of new materials for the formulation and manufacture of paints and coatings that simultaneously exhibit efficiency, are economical, and meet requirements that seek to reduce pollution.

Researchers specializing in painting technology and materials science in general, the academic community that participates in postgraduate teaching tasks, and particularly the professionals who make up the productive sector surely have different expectations in reading this book. Editors want at least some of the above objectives to be achieved; thus, this book will have reached its purpose.

The introductory chapter, "Protection of Materials," describes aspects related to metallic corrosion and failures as well as the action of fire on several materials.

The section "Metallic Corrosion and Failures" includes four chapters; they describe innovative methods of surface treatment to control the kinetics of metallic corrosion.

The first chapter reports on the novel results obtained for high-performance silica-poly(methyl methacrylate) (PMMA) and silica-epoxy coatings, containing active corrosion inhibitors, such as cerium salts; also, it compares their properties with results reported for a variety of hybrid coatings prepared by different approaches and formulations.

Other chapters, since electrodeposition is a low-cost and low-temperature method both for producing metal matrix composite coatings and for attending the combined mechanisms of corrosion and erosion, describe monolayer and multilayer alloy systems deposited by applying this process (composite coatings through the codeposition of nanoparticles).

The following chapter reports a proposal for reducing the predominant asset of the investment cost for proton exchange membrane (PEM) electrolyzer by using protective and easily upscalable coating technologies. Vacuum plasma spraying, a versatile technology for applying various types of coatings to a wide range of surfaces, is described to produce highly stable and multifunctional coatings for cost-effective interconnectors of PEM electrolyzers.

The next chapter refers to the production of thin films (nanocoatings); it is a technological field with many applications to elaborate materials with new properties to be used as corrosion protection of traditional metals. Two hydrophobic corrosion coatings for possible use over aluminum heat exchanger geothermal power plants are described.

The section "Fire Action on Materials" includes a chapter in which fire retardant coatings are studied; this issue is very important since fire causes human and economic loss. Fire retardant coatings are often required to protect a wide range of both flammable and nonflammable products against fire. The degree of fire retardancy mainly depends on the efficiency of formulations as well as on coating thickness, the type of substrates, etc. The wood behavior protected by using fire-retardant coatings is explained.

Finally, the editors want to point out that they hope that this book will become a reference work that offers a satisfactory answer to the questions related to painting technology for researchers, postgraduate teachers, and specialists in the production sector.

Carlos A. Giudice
National Technological University, La Plata Regional Faculty
Argentina

Guadalupe Canosa
Research and Development Center in Coatings Technology
National Technological University, La Plata Regional Faculty
Argentina

Section 1

Introduction

Chapter 1

Introductory Chapter: Protection of Materials

Guadalupe Canosa and Carlos Alberto Giudice

Additional information is available at the end of the chapter

http://dx.doi.org/10.5772/intechopen.69853

1. Introduction

The science and technology of materials have as objective provided access to specific knowledge in this field and besides generated matters of interest for the development of original researches leading to new materials and, as a consequence, to proper protection methods to achieve high economic impact.

Since iron and steels constitute a group of widely used materials in civil and industrial construction, this chapter describes firstly the fundamental concepts of metal corrosion and the most frequent failures originating on service. Besides, in this chapter, it was considered of interest to include aspects inherent to the physicochemistry of fire and its mechanism of spreading because of the significant human and material losses produced year after year by action of the fire.

It is appropriate to mention that the different chapters of this book describe innovative methods of surface treatment to control the kinetics of metallic corrosion and the action of fire on several materials.

2. Metallic corrosion and failures

Accidents arising from the *metallic corrosion* can produce injury or death of people by explosion, fire, and so on. The economic losses are classified into direct and indirect; the first includes the replacement of corroded materials, labor, periodic maintenance (coatings, cathodic protection, inhibitors in closed circuits, etc.) while the last involve aspects such as the discontinuity in the productive system, the loss and the contamination of raw materials and finished goods, and so on. The indirect losses are usually between 8 and 10 times the direct ones.

In industrialized countries, the total economic losses reach values between 3.5 and 4.5% of gross national product, despite applying all available technologies. It should also be mentioned that the frequency of failure in metals by the various forms of corrosion reaches an average level of 60% (the remaining 40% is due to mechanical failures).

Metallic corrosion is usually defined as the destructive attack of a metal by chemical or electrochemical reaction with the environment [1]. Chemical corrosion involves the alteration of a metal in a non-ionic medium, such as gases or non-condensed vapors, high temperature, and so on. On the other hand, the electrochemical corrosion occurs with a simultaneous transport of electricity through the metal and the electrolyte (saline, atmosphere, seawater, etc.).

The most common metallic corrosion takes place electrochemically; it requires, as it is already put, electrical conductivity. Metals are electronic conductors of first specie while solutions and pure liquids are electrolytes of second specie.

Metallic or electronic conductors transport electricity through the electrons. The metals consist of a relatively rigid network of positive ions and of mobile electrons. When an electrical potential is applied, the electrons move in one direction while the positive ions remain static; the quoted electricity transport is produced without appreciable movement of matter. Since the electrons have a negative charge, the direction in which they move is the opposite at which is conventionally considered as positive current.

Meanwhile, the electrolytes carry the electric current through ions, that is, with a significant movement of matter. Ions are atoms or groups of atoms that have lost or gained electrons, reaching in this way positive charges (loss of electrons) or negative (gain of electrons), **Figure 1**. The positive ions (cations) move in the direction of current and the negative ions (anions) in the opposite one.

The determinant factors of metallic corrosion are the heterogeneity of metal (phases in alloy, remainder mechanical stresses, etc.) and/or of electrolyte (gradients of concentration, differential aeration, etc.). Meanwhile, the chemical nature of electrolyte (ion conductivity or equivalent) significantly influences the kinetics of the corrosive process and the geometry of the corrosion cell (higher conductivities usually favor the location of electrodes more distant from each other than solutions of high resistivity).

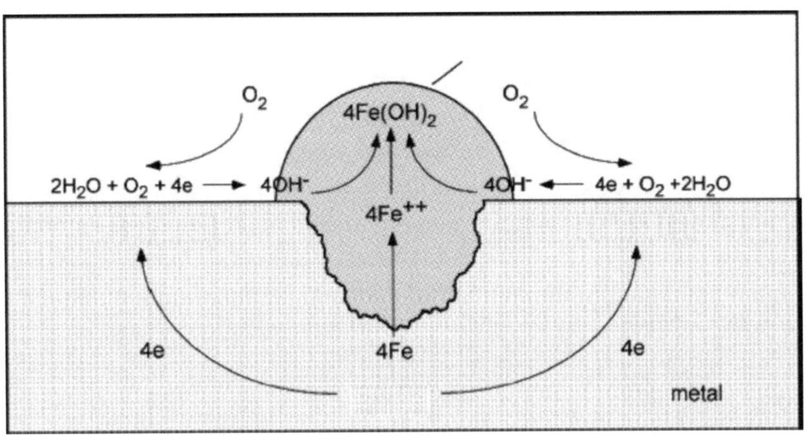

Figure 1. Corrosion mechanism.

A measure of the electrochemical kinetics (rate of reaction on electrode) is given by the equation $i = z\ F\ V$, where i is the current density (current per unit area of electrode), z the number of equivalents per mole, F the Faraday constant (96,500 Coulomb/equivalent), and V the rate of reaction in moles per unit area and time.

The abovementioned heterogeneity leads at the metal-solution interface to a gradient of electric potential between two adjacent areas. From a thermodynamic point of view, the quoted potential gradient is correlated with a difference of free energy ΔG. This is a thermodynamic function that is used as a criterion of spontaneity; it depends only on the initial and the final states, that is, that it is independent of the path: it decreases its value in a spontaneous transformation, either of physical or of chemical nature at constant temperature and pressure. Accordingly, it is possible to conclude that metal surfaces with high free energy are thermodynamically unstable and therefore tend to spontaneously evolve into a state of lower energy and greater stability.

The free energy is related to the electromotive force of a corrosion cell through the equation $\Delta G = -\ z\ F\ E$, where E is the reversible potential in volt, z the number of equivalent by mol, and F the Faraday constant.

Consequently, ΔG is the electrical work carried out by corrosion cell; it is observed that a reaction occurs spontaneously, at a constant temperature and pressure, when the value of E is positive.

Electrochemical corrosion is actually a network of shorted galvanic cells arranged on the metallic surface. Metal dissolves in the anode areas in equivalent relation to the reaction that takes place in the cathodic areas. In general, the anodic reaction is faster in almost all media, that is, that cathodic reaction is usually the decisive stage of the overall speed of corrosion process.

The *cathodic reaction*, in deaerated solutions, involves the reduction of protons (fast in acid and slow in neutral and alkaline media); instead, the quoted reaction, in aerated solutions, is accelerated by the reduction of dissolved oxygen:

$$2\,H^+ + 2\,e \leftrightarrow H_2 \tag{1}$$

$$O_2 + 2\,H_2O + 4\,e \leftrightarrow 4HO^- \tag{2}$$

In both cases, there is an alkalization of the cathodic area, either by the decrease of the concentration of protons or directly by the generation of hydroxyl groups.

Meanwhile, the *anodic reaction* involves loss of electrons, from atoms of higher free energy, arranged on the metallic surface:

$$Fe^0 \leftrightarrow Fe^{+2} + 2\,e \tag{3}$$

$$Fe^0 \leftrightarrow Fe^{+3} + 3\,e \tag{4}$$

As a result of the ferrous and/or ferric ions reacting with ions hydroxyl of the medium for generating hydroxides, acidification of the anodic area occurs.

The sum of the anodic and cathodic hemi-reactions, in aerated media, is as follows:

$$Fe^0 + H_2O + 1/2\,O_2 \leftrightarrow Fe(OH)_2 \quad (5)$$

$$2\,Fe^0 + 3\,H_2O + 3/2\,O_2 \leftrightarrow 2Fe(OH)_3 \quad (6)$$

The reaction in a corrosion cell involves the formation of hydrated ferrous oxide (ferrous hydroxide), which forms a first barrier for the diffusion of oxygen (polarization). This hydroxide is white in its pure state and has a pH of 9.5 in saturated solution. In a second sequential reaction, the hydrated ferrous-ferric oxide, which is the intermediate layer, is formed. This product, black in color, has magnetic properties. Subsequently, the reaction leads to the generation of hydrated ferric oxide, which makes up the third (external) layer of the oxidized system. This compound is orange/dark red in color and has a nearly neutral pH in saturated solution; exists as αFe_2O_3 (non-magnetic, with higher free negative energy of formation, i.e., more stable) and as γFe_2O_3 (magnetic).

On the other hand, the failures of metals take place by different causes due to the great amount of variables involved; as previously mentioned, the frequency of failure in the metals by corrosion reaches average levels of 60% in the different productive sectors. The types of corrosion failure and their frequency are given in **Table 1**.

Uniform corrosion. It is characterized in that the cathodic and anodic areas are modified alternately in space and time; as examples, it is possible to cite the case of a metal in direct contact with a solution of reduced electrical conductivity (the corrosion products, due to the reduced distance between the electrodes, are deposited simultaneously on the anodic and cathodic areas controlling the kinetics of process) and also the case in which the metal is exposed to

Type of failure	Failure frequency, %
Uniform corrosion	31.2
Corrosion-fatigue and corrosion under tension	23.4
Corrosion by pitting	15.7
Inter-granular corrosion	10.2
Corrosion-erosion, corrosion-wear and corrosion-cavitation	8.4
High-temperature corrosion	2.3
Corrosion by welding	2.1
Thermo-galvanic corrosion	2.0
Galvanic corrosion and corrosion in concentration cells	1.4
Corrosion by electrolysis	1.1
Corrosion by selective attack	1.0
Microbial corrosion	0.7
Corrosion by hydrogenation	0.5

Table 1. Types of failure and frequency.

high temperature in a relatively dry atmosphere. Preventive measures generally include selecting suitable materials for each aggressive medium, changing or inhibiting the electrolyte (closed systems), specifying resistant coatings, and designing anodic protection (passivation).

Corrosion fatigue. It is characterized by the action of alternating tensions in the presence of a corrosive medium. The causes are basically the same that can be attributed to static fatigue but adding cyclic loads. The deteriorating effect of combined fatigue and corrosion is much greater than the sum of individual damages. The most suitable measures to avoid this type of corrosion are to eliminate the cyclic tensions, increase the size or thickness in critical sections, reduce the concentration of stresses or redistribute them, provide sufficient flexibility to diminish over-fatigue by thermal expansion, control the vibration or shocks, eliminate the sudden changes in loads, temperature, or pressure, specify the right surface finishing, and select the appropriate protective system.

Corrosion under tension. It consists of premature breakage caused by the combined action of corrosive medium and residual or applied stress on the piece of metal, that is, that it takes place by combining high efforts and the presence of an electrolyte. Efforts by static charges in the metal surface and corrosive action that diminishes the section of the piece may exceed the elastic limit and even the breaking load. The forms of controlling this failure are to reduce mechanical tensions, ensure a sufficient flexibility, increase the size of the critical sections, select materials in the joints with a similar expansion coefficient, design adequate protection, and use a medium of suitable nature and composition.

Corrosion by pitting. It is a localized phenomenon that produces an appreciable penetration in the metal, generating either cavities or a discontinuity of the protective coating that lead to the formation of a concentration cell. To avoid this pathology, it is convenient to control the properties and the main characteristics of protective film (dry and wet adhesion, thickness, permeability, etc.), select a good geometry to prevent attacks, and specify properly the electrolytic medium.

Inter-granular corrosion. It is the preferred attack on grain boundaries of a metal or an alloy; it is characterized by a selective deterioration and an inter-crystalline cracking along inter-granular streaks (e.g., in stainless steels in chrome-deprived areas). Frequently, the specifications contemplate to select materials with a suitable thermal treatment for each particular case and realize weldings that do not generate temperatures superior to those used in the pretreatment of material.

Corrosion-erosion. The failure generated by the relative movement of the electrolytic medium (generally accelerated by abrasion due to the presence of solid particles in suspension) releases the corrosion products adhered to metal (depolarization) and also causes surface wear. For satisfactory corrosion-erosion control, it is appropriate to decrease the fluid velocity to achieve laminar movement, suppress the localized turbulence and the discontinuous flows, eliminate the abrupt changes in the direction of flow (aligning sections of ducts), avoid the obstructions, increase the material thickness in critical areas, design anodic parts so they can be changed quickly, specify the surface roughness, select the suitable coatings, and carry out cathodic protection.

Corrosion wear. It is defined as the deterioration located at the interface between two surfaces in contact, accelerated by a relative movement of sufficient amplitude to produce slippage. Generally, it occurs under heavy loads and instantaneous movements produced by high-frequency vibrations; the wear of surface-protective film (inorganic primers, organic coatings, etc.) can initiate a corrosion process. The main prevention methods to avoid corrosion wear are to eliminate the transmission of vibrations, introduce barriers between metals that slip, increase the load to slow the movement, provide protective layers to porous materials or use suitable lubricants, isolate those moving parts of the static ones, and finally increase the abrasion resistance.

Corrosion cavitation. It is associated with vapor bubbles arranged inside the liquid that collapse on the surface of the solid. Repeated collapses on a metal surface can deteriorate the protective film and severely deform the surface, fracturing it or generating fatigue. Low-pressure areas are created by divergent flows, vibrations, and so on. To control these damages, it is very important to select conditions that diminish absolute pressure, reduce hydrodynamic pressure differences, control the vibration, design the system to avoid formation or accumulation of bubbles, prevent the entry of dispersed air, select resistant materials or coatings, specify the finishing polishing, use cathodic protection, and so on.

Corrosion by high temperature. It is associated to the effect of atmospheric conditions and the presence of gases, metals, and/or molten salts at high temperature; the kinetics depends on the nature of the metals, the composition of the medium, and the time of exposure. The reduced dimensional stability of the corrosion products (hydration/dehydration by thermal changes) produces tangential cutting stress to the surface leading to the partial detachment of the different oxide layers, generating heterogeneities that favor corrosive processes. The most recommended therapies are to select materials stable to the thermal action, adjust the nature and/or composition of the medium, and regulate, if possible, the contact time.

Corrosion by welding. A weld can have low corrosion resistance due to the chemical nature of the electrode (e.g., it should be used with those having a low hydrogen content), to the residual stress and to the metallurgical structure of the weld zone. Corrosion in welding joints can be avoided by careful selection of materials, of the technique used, and of the type of finishing.

Thermo-galvanic corrosion. It is the result of the operation of a galvanic cell generated from a temperature gradient; the heating and the heat dissipation in heterogeneous form are the responsible factors for the formation of this cell. The most efficient actions are to avoid point heating and/or unequal cooling, use a continuous and adherent coating, and introduce thermostated components from the outside to the system.

Galvanic corrosion. It involves the corrosion associated with the current resulting from the contact of different electrodes (metals of dissimilar chemical nature) arranged in a conducting electrolyte that closes the circuit of the cell. The most important preventive measures are to eliminate interaction of diverse metals or to produce a complete dielectric insulation, avoid contact of a small anode and a large cathode, extend the distance between dissimilar metals in conductive media, design the anodic parts that can be easily replaced or apply thicker protective films, use suitable protective systems and regulate the degree of aeration, temperature, composition, or movement of the medium that is suitable for the metal coupling.

Corrosion by concentration cells. It is made up of a galvanic cell in which the electromotive force is due to the concentration difference of one or more reagents. The main causes are given either by differential aeration (different partial pressure of oxygen) generated in cracks, adherent deposits, and deep depressions that influence the diffusional process of oxygen and the existence of gradients of concentrations in the electrolyte generated by different causes. The most effective measures are to reduce surface irregularities especially in areas of heat transfer or where chemical reagents or oxygen are introduced, design drainage and a uniform environment, select forms that allow easy cleaning and application of protective layers, remove solids in suspension by filtration, use continuous welds, suppress porosity and cracking, and eliminate fibrous and/or absorbent packings.

Corrosion by electrolysis. It is generated by a current flow, that is, electric currents generally of an alternating nature, which cannot be controlled; they are often originated by sources external to the structure (e.g., bad ground connections, etc.), which enter through a conducting medium. It is convenient to connect properly the equipment to ground, isolate the apparatus from structures, use non-conducting fluids, eliminate errant or vagabond current sources, and incorporate sacrificial (cathodic protection) plates in the anodic areas near insulation joints.

Corrosion by selective attack. It is based on a process of extracting a soluble component from an alloy; generally, the percolation of the alloy occurs by the action of a solvent on an element of the metal (e.g., zinc, aluminum, etc.), which separates and consequently generates a corrosive action. The most appropriate measures involve selecting materials suitable for performing efficiently in the electrolytic medium in which the part or structure is inserted, reducing the aggressiveness of the medium if feasible (e.g., in closed systems), and using suitable protection methods.

Microbial corrosion. Bacteria and fungi, individually or together, and the subproducts of the biological activity attack the metal and/or the coating. The mentioned products (e.g., organic and inorganic acids and alkalis) display a significant aggressiveness to materials. Consequently, considering the causes described, it is convenient to avoid contamination, use specific biocides, control chemically the environment, select properly the protective coatings, and clean the surfaces as often as necessary.

Corrosion by hydrogenation. It is manifested by the reduction of the mechanical resistance produced by the inclusion of hydrogen gas in the crystal structure of the metal. The most common causes are linked to an inadequate de-oxidation and, fundamentally, to an oversizing of the cathodic protection. The most suitable therapies are to perform a suitable surface preparation, select properly coating systems, induce compressive stresses, heat the metallic substrate to 90–150°C, and systematically control the electrical potential of the metal substrate modified by the cathodic protection.

It is worth mentioning that coating systems are the most convenient methods for controlling the kinetics of metallic corrosion from a technical-economical viewpoint [2–6].

3. Fire action on materials

Fire is an energetic manifestation that constantly accompanies human activity; therefore, the emerging risk must be assumed.

Fire develops strongly exothermic chemical reactions, starting when oxidizer and combustible are in a sufficient energetic state (activation energy). The combustible includes substances that are not in their maximum oxidation state; in general, any material containing carbon and/or hydrogen can be oxidized and therefore be combustible. The most important oxidizer is air, which is composed in its fifth part by oxygen; during combustion, the other components remain unchanged (except at very high temperatures) and accompany the products of combustion in the fumes. Part of the energy released in the reaction is dissipated, generating an increase in the temperature of the medium and the remainder is transferred to the reaction products providing the activation energy for the process to continue; if this is not enough, the combustion stops.

The knowledge of the physicochemical theory of combustion has allowed the development of products and systems of defense against fires. Nevertheless, the losses occasioned continue to be one of the greatest tragedies of modern civilization. Taking into account the current technology of fire-retardant treatments (impregnation, coatings, etc.), it is important to mention the generic concept of "passive protection against fire," in which the efficiency is independent of human activity.

The research and development studies are thus significant to reduce the combustibility of materials and the speed of propagation of the flame front as well as to keep during the conflagration the mechanical properties of structures based on either combustible and non-combustible materials. The design of the constructions also plays a very important role.

The true magnitude of the fire problem is remarkable when considering the human and material losses occurring year after year. Thus, for example, 25% of the deaths caused by fire are due to people remain trapped inside buildings; the majority of victims are younger than 10 or older than 70 years. Considering accident deaths, those caused by the fire action are only surpassed by the car crashes.

With regard to economic losses, they reach in many countries a value nearly to 0.25% gross national product. Fire generates significant problems in civil constructions, ships, offshore structures and industrial plants; in many cases, the use of both untreated materials and conventional coatings contribute to the fire spreading.

Often, there are also significant indirect losses of difficult evaluation such as the decrease of income by the total or partial interruption of the activity of a company, the decrease in customers, the increase of replacement costs of installations and equipment, and so on. It is estimated that out of every five companies that have had a major fire, four of them disappear within three years of the incident.

In relation to cultural heritage and historical buildings, material losses are remarkable. For example, the Argentinian Theater of La Plata (Buenos Aires, Argentina) was completely destroyed by a fire and the Theater della Scala (Venice, Italy) was seriously affected by another conflagration, in the decades of the 70 and 90 of the previous century, respectively.

As a consequence of the spectacular fires in historic and massive concurrence buildings, many countries adopted regulations for the control of materials flammability. The latter led to developments of intrinsically fire-resistant materials, retardant treatments, and a large number of test methods to evaluate the reaction to fire of the materials. It is also important to mention that for many years now, insurance companies have found that the way to deal with fire is through the prevention and the use of fire-proofing materials.

Stability of construction materials. The fire action on construction materials is significant; thus, for example, calcareous collapses rapidly by dilation and by contraction during drying.

Concerning the concrete, it exhibits satisfactory response to high temperatures if perfectly anchored. For its part, reinforced concrete presents adequate behavior up to 300–330°C if its aggregates are small in size; the iron framework begins to lose resistance when reaching a critical temperature of 500–550°C.

As regards gypsum, it is gradually dehydrated above 120°C and up to 180°C, loosing cohesion at 700–800°C.

The load-bearing iron and steel structures (made by forging or rolling) are plastically deformed by the action of heat, essentially when the pressure leads to lose their static equilibrium; at approximately 500°C, these materials halve their structural strength.

Wood and wood products were widely used in the construction of historic buildings; in spite of behaving like combustible materials and to be vulnerable in cases of fire, in general they display a considerable fire resistance (small decrease of area attributable to the low thermal conductivity of the superficially formed carbonaceous layer). Untreated wood begins to burn at 300°C but that treated with suitable fire retardants does not release so much smoke (the gases are non-toxic and non-combustible). The losses in cases of conflagration are always lower than in the constructions with iron and other metals and, once the origin of the fire has been eliminated, the wood is characterized by exhibiting a behavior corresponding to a self-extinguishing material.

All the abovementioned values have a singular meaning, since the average temperature of the fire ranges usually between 700 and 800°C.

Fire spreading. The speed of propagation of the flames plays a preponderant role in the advance of the fire front; as mentioned, the toxicity of gases and fumes is a significant variable. The room propagation involves the three forms of heat transfer (convection, radiation, and conduction): in the interior of a building, by conduction through the walls when the thermal insulation is reduced or by convection when there are open stairs while between adjacent buildings by radiation through the openings, doors, and windows.

It is worth mentioning that in the buildings under construction, expansion, or demolition, the probability of fire is particularly high during (i) the heating, welding, and cutting processes; (ii) the transport of flammable liquids and materials; and (iii) the use of electrical equipment with precarious installations.

Total thermal load and fire load. It is significant to determine the degree of risk and adequate security measures, particularly for civil buildings designed to permanently or transiently accommodate a large number of people (schools, libraries, hospitals, hotels, restaurants, auditoriums, theaters, cinemas, shops, etc.) and industrial units built to store and/or manufacture products, equipment, and appliances (petrochemicals, automotive terminals, medical laboratories, sawmills, etc.).

The total combustion risk of a building is calculated by considering the caloric content of the building (fire load including the building itself) and the enthalpy level of the content (fire load involving human lives and properties). The Pourt method was developed from the value of fire load and is widely used to determine the total risk of buildings; the fire charge density, calculated by dividing the fire load by the building surface, is also a widely considered variable.

Performance of coatings in fire. Coatings in particular and coating systems in general play well-defined actions against fire action [7–19]; they may

- Promote the spreading. Generally, conventional coatings have a low ignition point, so by thermal action they release combustible gases, which ignite and release caloric energy; the last one in turn becomes the energy of activation that promotes the spreading of conflagration front.
- Display inertia. Some commercial products of reduced efficiency do not alter the fire performance of the bare substrate or only achieve a limited retarding action.
- Delay evolution or extinguish fire. The retarding effect interrupts, in one or more stages, the combustion; the process ends in an acceptable lapse, often before the ignition takes place, **Figure 2**.

Testing methods. The analysis of the current regulations in the world indicates the existence of a great number of tests of different characteristics to determine the reaction, the resistance, and the stability against fire of the constructive elements. The results depend on the type and shape of the specimen, the intensity and time of action of the external energy source, and so on.

Figure 2. Left, panel without treatment and right, film of intumescent coating, both after the fire action.

The main variables considered include the size and position of specimen, the type and magnitude of energy source, the way and rate of heating, the duration of test, and valued indices; the fire performance varies according to the method applied.

In many occasions, the abovementioned is a technological barrier for the export/import of either fire retardants or treated materials. A political decision must be taken to impose common test methods at least at the regional or continental level, resulting in adequate reproducibility, that is, that in the case of operators working in different laboratories or in the same laboratory at different times, achieve comparable individual results (low dispersion of the mean value) by using the same method on an identical material.

Acknowledgements

The editors are grateful to the National Technological University (UTN) and the National Council of Scientific and Technical Researches, Argentina (CONICET).

Author details

Guadalupe Canosa[1,2] and Carlos Alberto Giudice[2]*

*Address all correspondence to: giudiceca@gmail.com

1 Centro de Investigación y Desarrollo en Tecnología de Pinturas, CIDEPINT (CICPBA-CONICET), La Plata, Argentina

2 Centro de Investigación y Desarrollo en Ciencia y Tecnología de Materiales, CITEMA (UTN-FRLP), Berisso, Argentina

References

[1] Giudice C, Canosa G. Protección de materiales metálicos. Editorial edUTecNe, Argentine; 2016. pp. 1-284, ISBN 978-987-1896-71-4 (online), ISBN 978-987-1896-70-7 (printed). Available from: http://www.edutecne.utn.edu.ar/pinturas_recubrimientos/proteccion_materiales_metalicos.pdf

[2] Dimitirev Y, Ivanova Y, Iordanova R. History of sol-gel science and technology (review). Journal of the University of Chemical Technology and Metallurgy. 2008;**43**(2):71-97

[3] Kahraman MV, Kuğu M, Menceloğlu Y, Kayaman-Apohan N, Güngör A. The novel use of organo alkoxy silane for the synthesis of organic–inorganic hybrid coatings. Journal of Non-Crystalline Solids. 2006;**352**(21-22):2143-2151

[4] Mafi R, Mirabedini S, Naderi R, Attar M. Effect of curing characterization on the corrosion performance of polyester and polyester/epoxy powder coatings. Corrosion Science. 2008;**50**(12):3280-3286

[5] Pascault J, Williams R. Epoxy Polymers. Weinheim: Wiley-VCH Verlag GmbH & Co, KGaA; 2009. pp. 1-357. ISBN: 978-3-527-32480-4

[6] Thuy D, Xuan H, Nicolay A, Paint Y, Olivier M. Corrosion protection of carbon steel by solvent free epoxy coating containing hydrotalcites intercalated with different organic corrosion inhibitors. Progress in Organic Coatings. 2016;**101**:331-341

[7] Bahlakeh G, Ramezanzadeh B, Reza Sae M, Terryn H, Ghaffari M. Corrosion protection properties and interfacial adhesion mechanism of an epoxy/polyamide coating applied on the steel surface decorated with cerium oxide nanofilm: Complementary experimental, molecular dynamics (MD) and first principle quantum mechanics (QM) simulation methods. Applied Surface Science. 2017;**419**:650-669

[8] Bardon J, Apaydin K, Laachachi A, et al. Characterization of a plasma polymer coating from an organophosphorus silane deposited at atmospheric pressure for fire-retardant purposes. Progress in Organic Coatings. 2015;**88**:39-47

[9] Dasari A, Yu Z, Cai G, et al. Recent developments in the fire retardancy of polymeric materials. Progress in Polymer Sciences. 2013;**38**(9):1357-1387

[10] Gu J, Zhang G, Dong S, et al. Study on preparation and fire-retardant mechanism analysis of intumescent flame-retardant coatings. Surface and Coating Technology. 2007;**201** (18):7835-7841

[11] Han Z., Fina A., Camino G. Organosilicon Compounds as Polymer Fire Retardants. In: Papaspyrides C and Kiliaris P (eds.). Polymer Green Flame Retardants. Elsevier, Amsterdam, Netherland. 2014; 389-418.

[12] Han Z, Fina A, Malucelli G. Thermal shielding performances of nano-structured intumescent coatings containing organo-modified layered double hydroxides. Progress in Organic Coatings. 2015;**78**:504-510

[13] Hu Y., Yu B., Song L. 3 - Novel fire-retardant coatings. In: Novel Fire Retardant Polymers and Composite Materials. Edited by: De-Yi Wang. Elsevier, Amsterdam, Netherland. 2017; 53-91.

[14] Li G, Han J, Lou G, et al. Predicting intumescent coating protected steel temperature in fire using constant thermal conductivity. Thin Walled Structures. 2016;**98**(A):177-184

[15] Li H, Hu Z, Zhang S, et al. Effects of titanium dioxide on the flammability and char formation of water-based coatings containing intumescent flame retardants. Progress in Organic Coatings. 2015;**78**:318-324

[16] Liu Z, Dai M, Zhang Y, Gao X, Zhang Q. Preparation and performances of novel waterborne intumescent fire retardant coatings. Progress in Organic Coatings. 2016;**95**:100-106

[17] Pereyra A, Canosa G, Giudice C. Nanostructured protective coating systems, fireproof and environmentally friendly, suitable for the protection of metallic substrates. Industrial Engineering Chemistry Research. 2010;**49**(6):2740-2746

[18] Ullah S, Ahmad F, Shariff A, Bustam M, Gonf G, Gillani Q. Effects of ammonium polyphosphate and boric acid on the thermal degradation of an intumescent fire retardant coating. Progress in Organic Coatings. 2017;**109**:70-82

[19] Chen W-Y, Wang Y-Z, Chang F-C. Thermal and flame retardation properties of melamine phosphate-modified epoxy resins. Journal of Polymer Research. 2004;**11**(2):109-117

Section 2

Metallic Corrosion and Failures

Chapter 2

Organic-Inorganic Hybrid Coatings for Corrosion Protection of Metallic Surfaces

Samarah V. Harb, Andressa Trentin,
Ruben F. O. Torrico, Sandra H. Pulcinelli,
Celso V. Santilli and Peter Hammer

Additional information is available at the end of the chapter

http://dx.doi.org/10.5772/67909

Abstract

A variety of organic-inorganic hybrids have been designed to act as anticorrosive coatings of metallic substrates. Among them, epoxy-silica and poly(methyl methacrylate) (PMMA)- silica hybrids, prepared by the sol-gel process and deposited onto steel or aluminum alloys, have demonstrated high anticorrosive efficiency combined with high thermal and mechanical resistance. Lignin, carbon nanotubes, and graphene oxide have been incorporated into PMMA-silica hybrids as reinforcement agents, and cerium (IV) as corrosion inhibitor. Both hybrids were characterized in terms of their structural and thermal characteristics using different pectroscopies, microscopies and thermogravimetric analysis. Both hybrids present homogeneous nanostructure composed of highly condensed silica nanodomains covalently bonded to the polymeric phase. The transparent coatings with a thickness of 2–7 μm have low surface roughness, high adhesion to metallic substrates, elevated thermal stability, and excellent barrier behavior. Electrochemical impedance spectroscopy showed for coated samples a high corrosion resistance of up to 50 GΩ cm2 and durability >18 months in saline solution. Further improvement of corrosion resistance, thermal and mechanical stability was achieved by incorporation of lignin, carbon nanotubes, and graphene oxide into PMMA-silica matrix, and a self-healing effect was observed after Ce(IV) addition. The results are compared and discussed with those recently reported for a variety of hybrid coatings.

Keywords: organic-inorganic hybrids, anticorrosive coatings, corrosion inhibitors, nanocomposites, self-healing, sol-gel process

1. Introduction

Since the discovery of copper in prehistoric times, an extensive diversity of metallic materials has emerged, as pure metals or metal alloys. Brass, bronze, steel, titanium, and aluminum alloys are currently the most applied metallic materials, notwithstanding the natural tendency to suffer corrosion under aggressive conditions and thus return to their original ore. To overcome issues related with economical losses and lack of safety, occasioned by metal corrosion, several protection methods have been developed, including the use of:

new alloys with higher corrosion resistance, but in addition to the high cost associated with their development, the use of new alloys requires the replacement of the metallic components;

corrosion inhibitors, substances which reduce or even eliminate corrosion, when present in suitable concentrations in the corrosive medium. Inhibition is accomplished by one or a combination of several mechanisms, such as adsorption, forming a ultrathin film with a thickness of only few molecular layers; in form of visible bulky precipitates, which coat the metal surface; or other common methods consisting of the combination of adsorption, conversion, and oxidation processes to form a passive layer. Some examples of most applied inhibitors are phosphates, chromates, silicates, hydroxides, carbonates, sulfates, aldehydes, amines, nitrogen heterocyclic compounds, urea, among others [1];

cathodic protection that uses a sacrificial metal to protect the metallic structure of interest. It is commonly used to prevent corrosion in large port structures, offshore platforms, and pipelines to transport water, oil, and gas;

conversion layer produced by converting the metal surface into a corrosion-resistant form. The main processes include anodizing, phosphatizing, and chromating, and they are frequently used as pretreatment for subsequent overcoats [1]. Anodizing is based on the formation of a protective surface layer, formed by oxides and hydroxides, by application of an external current. In an electrochemical cell, the surface of the metal anode is transformed into an oxide layer of defined thickness, which improves significantly the corrosion resistance and the adhesion of subsequent paints. This method is frequently applied to protect aluminum alloys but can be used also for titanium, zinc, magnesium, and other metal substrates. Besides the presence of microscopic fissures in the anodizing coating that can lead to corrosion, another drawback is the susceptibility of the oxide layer to chemical dissolution in the presence of high- and low-pH environments. Phosphatizing is mostly applied on steel substrates to produce an insoluble and porous phosphate layer that serves as an excellent base for coatings. For instance, car bodies have been phosphatized prior to the application of coatings for many years [1]. Alternatively, surface passivation using chromate conversion coatings has been used especially to protect aluminum alloys in the aerospace industry, for metal fittings and for packaging steel [2, 3]. Such coatings are formed by the reduction of Cr(VI) species to hydrated Cr_2O_3; however, the conversion process as well as the final coating retains a small amount of unreacted Cr(VI), a highly toxic species which can be released to the environment. Recently, increasing efforts are focused on the development of innovative eco-friendly alternatives due to increasingly strict legislation regulations which demand a reduction of hexavalent chromates usage [3];

protective coatings applied on metal surfaces result in a barrier between the metal and the corrosive medium, thus preventing or minimizing the corrosion process.

The use of coatings on metallic surfaces has various advantages, such as relatively low costs, environmental compatibility, and the possibility to apply them on metallic components already in use. Consequently, different kinds of protective coatings have been developed, comprising metallic, inorganic, organic, or organic-inorganic materials. The application of many metal coatings, such as chromium, zinc, nickel, aluminum, and copper, involves usually inherent pollution and toxicity-related problems. The most widely used metallic coating is zinc, commonly deposited on carbon steel by hot-dip on a molten zinc bath, process called galvanization, after which the metal substrate acquires a zinc-rich top layer with a thickness of approximately 10 μm. Inorganic coatings comprise ceramics (silica, titania, zirconia, alumina), glass, carbon, etc [1]. Although the inorganic coatings present higher corrosion resistance compared to bare substrates, they usually exhibit residual porosity and stress-induced cracks, which limit their use as efficient corrosion barrier as they allow the diffusion of corrosive species to the underlying metal [4, 5]. Organic materials such as epoxy, poly(methyl methacrylate) (PMMA), polyurethane (PU), polyesters, fluoropolymers, and related paints, combined with anticorrosive primer containing various types of pigments, are widely applied as protective coatings. This is justified by the simplicity of deposition, their dense and homogeneous structure, and consequently high corrosion resistance in aggressive environments. However, their lack of thermal stability, mechanical resistance and adhesion to metallic surfaces can result in serious restriction of their long-term stability.

Organic-inorganic hybrids stand for a class of materials formed by the combination of a polymeric and a ceramic phase, resulting in a nanocomposite material with unique properties. New functionalities result from the synergy of both components, achieved by a careful adjustment of the nature, proportion, and the type of interaction at the interface of both phases. One of the most used methodologies to produce organic-inorganic hybrid materials is the sol-gel process, which allows due to its versatility to control the structure and the functional properties. Through hydrolysis and condensations reactions, the sol-gel route allows the obtain high purity, homogeneous, and structurally tuneable materials, which have a wide range of applications such as catalysts, drug release systems, photochromic devices, biosensors, transparent insulating films, and anticorrosive coatings with excellent barrier properties [6]. The latter characteristic is related to the possibility to prepare a dense organic-inorganic network structure by linking both phases covalently on the molecular scale, and furthermore, the ability to covalently bond the inorganic phase with metallic substrates, leading to highly adherent coatings. Consequently, intense research efforts are presently focused on the development of organic-inorganic hybrid coatings in form of passive barrier layers with low permeability for corrosive species such as chloride ions, water, and oxygen.

There are various methodologies to investigate the corrosion protection efficiency of coated metals; however, the most applied electrochemical techniques are electrochemical impedance spectroscopy (EIS), potentiodynamic polarization, and chronopotentiometry. Among them, EIS allows for a deeper analysis of the electrolyte/coating/substrate systems, due to the possibility to fit the data using equivalent electrical circuits, which permit to extract important electrochemical parameters such as coating capacitance, pore resistance, double layer capacitance, charge transfer resistance, water uptake, diffusivity, among others. Additional methods like salt spray test and immersion techniques are used according to different norms for the qualitative and quantitative evaluation of corrosion zones, pitting, and for the determination of corrosion rates. To evaluate the electrochemical performance of the protective system for a given corrosive environment and coating thickness, the most important criteria are (i) the magnitude of the initial impedance modulus obtained by EIS at low frequency, defined as corrosion resistance; (ii) the values of the open circuit potential, obtained by chronopotentiometry; and (iii) the time evolution of both parameters, to evaluate the long-term stability of the coatings. For industrial application, another important aspects have to be considered such as the simplicity of the synthesis process, low costs of reagents, and their environmental compatibility.

One example of an efficient corrosion protection of mild steel was recently reported for a hybrid system combining an epoxy-siloxane topcoat with an epoxy primer containing micaceous iron oxide and zinc phosphate pigments [7]. The electrochemical measurements showed a high-impedance modulus of up to 100 GΩ cm^2, remaining stable for more than 1 year in contact with 3% NaCl solution. The authors attribute the excellent protection to the high resistance of the coating against water uptake provided by suitable epoxy/primer combination and the relatively high thickness (~140 μm) of this coating system. In another recent study, Ammar et al. [8] report on high-performance hybrid coatings based on acrylic-silica polymeric matrix reinforced by SiO_2 nanoparticles, applied to mild steel with a thickness of 75 μm by brush coating. EIS measurements confirmed the high-corrosion protection efficiency with an impedance modulus of more than 10 GΩ cm^2, decreasing one decade after 90 days of immersion in 3.5% NaCl solution. Visuet et al. [9] obtained similar results for polyurethane/polysiloxane hybrid coatings containing TiO_2 as pigment. The EIS analysis showed that coatings loaded with 10-wt% TiO_2 (75 μm thick) were able to withstand 263 days, in 3.5% NaCl solution, with almost unaltered corrosion resistance of about 100 GΩ cm^2. Their model proposes that the TiO_2 pigment works as a charge (ionic) storage surfaces, thus enhancing the barrier property of the coating against electrolyte uptake.

The above results demonstrate that elevated anticorrosive performance is usually achieved for sophisticated barrier coatings with an average thickness in the order of dozens to hundreds micrometers. For the market, however, which aims on economic and efficient solutions, elevated thickness, and complexity of the coating system, implies elevated material costs and weight increase, issues that are hardly to be accepted, especially by the aerospace industry. In this regard, dos Santos and coauthors [10] have successfully prepared highly efficient PMMA-silica coatings having a thickness of only ~2 μm, which were able to withstand aggressive saline/acid (0.05 mol L^{-1} NaCl + 0.05 mol L^{-1} H_2SO_4) and 3.5% NaCl environments for up to 105 and 196 days, respectively, maintaining the corrosion resistance in the GΩ cm^2 range. The excellent performance of the primer free coating was explained by the high connectivity of reticulated sub-nanometric silica domains densely interconnected by short PMMA chain segments. Another results that confirmed the viability of thin hybrid films as efficient corrosion barrier have been reported in the study of Harb et al. [11]. The authors showed that the addition of cerium (IV) salt into PMMA-silica system results in a further improvement of the corrosion resistance and durability of the coatings applied to polished carbon steel by dip-coating. The electrochemical behavior of ~1.5 μm thick films reached for a Ce/Si molar ratio of 0.7% an impedance modulus of about 10 GΩ cm^2 (NaCl 3.5% solution) and remained stable within one order of magnitude for 304 days, a performance typically observed for high performance paint systems. The remarkable anticorrosive protection has been associated with the role of Ce(IV) as oxidation agent leading to an enhancement of the overall connectivity of the hybrid network, induced by the enhanced polymerization of organic and inorganic moieties.

In contrast to coating system designed as passive barrier, recent trends aim on the development of active multifunctional anticorrosive coatings with self-healing ability, high-thermal stability, and mechanical resistance, among other functionalities. Inspired by biological systems, the self-healing ability involves the complete recovery of the original properties of the material after suffering macroscopic lesions, induced by mechanical or chemical processes. Various strategies have been used to prepare self-healing coatings, usually containing an active compound, whether stored in microcapsules or incorporated into the coating. They can be activated by temperature increase, UV, pH gradient, breaking of capsules, or changes in the chemical environment [12, 13]. A number of studies report on the use of cerium salts (chloride and nitrates) and ceria nanoparticles as inhibitors, preventing corrosion by the self-healing ability in affected areas of inorganic, organic, and hybrid coatings. The resulting substantial lifetime increase is attributed to the formation of insoluble oxides and hydroxides in the corroded zones [3, 11, 14–17]. On the other hand, significant improvements of thermal and mechanical properties have been achieved by incorporation of clays, lignin, carbon nanotubes, graphene oxide, and graphene into polymeric or organic-inorganic matrices [18–20].

This chapter reports on recent results obtained for high-performance PMMA-silica and epoxy-silica hybrids coatings, correlating their structural properties with the corrosion protection efficiency, accessed by potentiodynamic polarization and electrochemical impedance spectroscopy. Moreover, several interesting finding are presented regarding PMMA-silica hybrids reinforced with lignin, carbon nanotubes, and graphene oxide to improve their thermal and mechanical properties, as well as some recent results on active corrosion inhibition by the self-healing ability of Ce(IV) containing PMMA-silica coatings.

2. Experimental

2.1. Epoxy-silica and PMMA-silica hybrid synthesis

All reagents used to epoxy-silica and PMMA-silica hybrids synthesis were purchased from Sigma-Aldrich and used as received, apart from the methyl methacrylate (MMA) monomer, which had been distilled before use to remove the polymerization inhibitor. The molecular structures of the epoxy-silica and PMMA-silica hybrid precursors are presented in **Figures 1** and **2**, respectively, and the synthesis procedures are summarized in **Figure 3**.

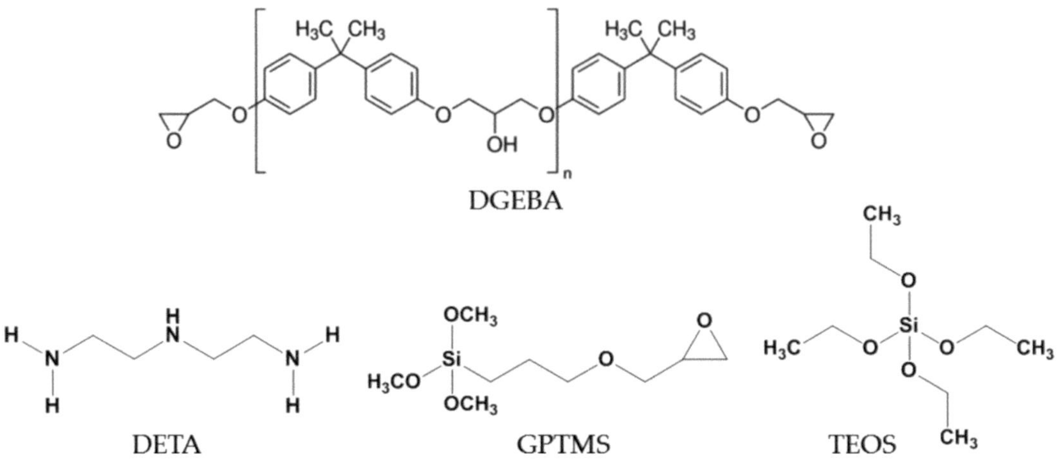

Figure 1. Molecular structures of the epoxy-silica hybrid precursors.

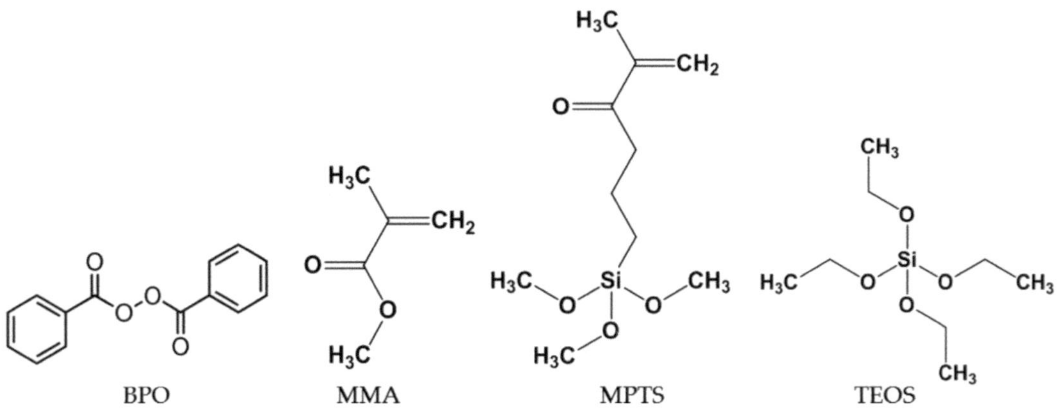

Figure 2. Molecular structures of the PMMA-silica hybrid precursors.

Epoxy-silica hybrids were prepared from the curing reaction of poly(bisphenol A-co-epichlorohydrin), glycidyl end-capped (DGEBA, Mn = 377 g/mol) with diethyltriamine (DETA) as hardener, and (3-glycidoxypropyl)methyltriethoxysilane (GPTMS), as coupling agent between the organic and inorganic phase, combined with the sol-gel hydrolysis and condensation reactions of tetraethoxysilane (TEOS) and GPTMS. In the first step, DGEBA and GPTMS were mixed with DETA in tetrahydrofuran (THF) solvent during 4 h at 70°C and 25 min at 25°C, under constant stirring in a reflux flask. In the next step, TEOS, ethanol, and acidified water (pH 1 using nitric acid) were added to the reflux system at room temperature and stirred for an additional 1 h. At this stage, the sol-gel reactions take place, as shown below, where the alkoxide precursors (TEOS and GPTMS) are hydrolyzed, forming Si–OH groups, Eq. (1), which subsequently condense with an initial alkoxide molecule, Eq. (2), or another Si–OH group, Eq. (3), yielding Si–O–Si bond and eliminating alcohol or water, respectively.

http://dx.doi.org/10.5772/67909

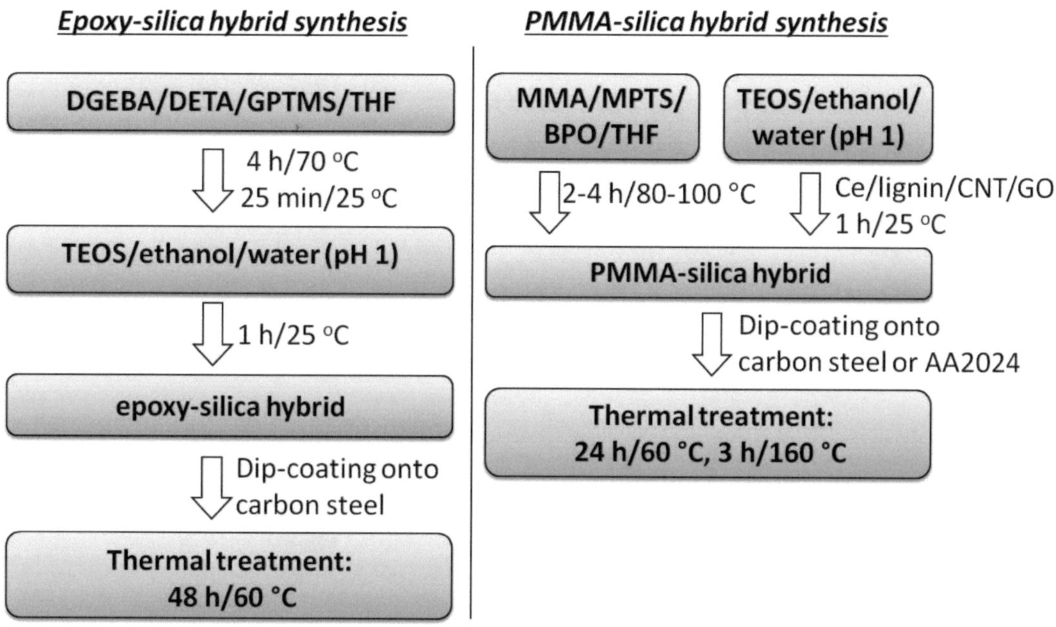

Figure 3. Synthesis procedures used to prepare epoxy-silica and PMMA-silica hybrids.

The homogeneous and transparent sols were used for the film deposition by dip-coating onto A1020 carbon steel.

$$\equiv\text{Si-OR} + H_2O \rightleftarrows \equiv\text{Si-OH} + \text{ROH} \quad (1)$$

$$\equiv\text{Si-OR} + \text{HO-Si}\equiv \rightleftarrows \equiv\text{Si-O-Si}\equiv + \text{ROH} \quad (2)$$

$$\equiv\text{Si-OH} + \text{HO-Si}\equiv \rightleftarrows \equiv\text{Si-O-Si}\equiv + H_2O \quad (3)$$

Two series of epoxy-silica hybrids were prepared, varying the amount of GPTMS or TEOS and keeping the molar concentrations of other compounds constant (**Figure 4**). In order to ensure a fully cured thermosetting, DETA was added in a proportion that resulted in one oxirane group for each hydrogen atom of the amine groups.

PMMA-silica hybrids have been prepared by the radical polymerization of methyl methacrylate (MMA) and 3-(trimethoxysilyl)propyl methacrylate (MPTS, also known as TMSM) using benzoyl peroxide (BPO), as thermal initiator of the polymerization, and tetrahydrofuran (THF) as solvent. The sol-gel route has been used to perform the hydrolytic condensation of tetraethoxysilane (TEOS) and MPTS, using ethanol and acidified water (pH 1 using nitric acid), during 1 h at room temperature. In the presence of acidified water, alkoxide precursors (TEOS and MPTS) are hydrolyzed and subsequently condensed to form Si–O–Si bonds. After mixing the organic and inorganic precursor, the obtained transparent and homogeneous sols were used to deposit few micrometer thick films onto A1020 carbon steel or AA2024 aluminum alloy substrates.

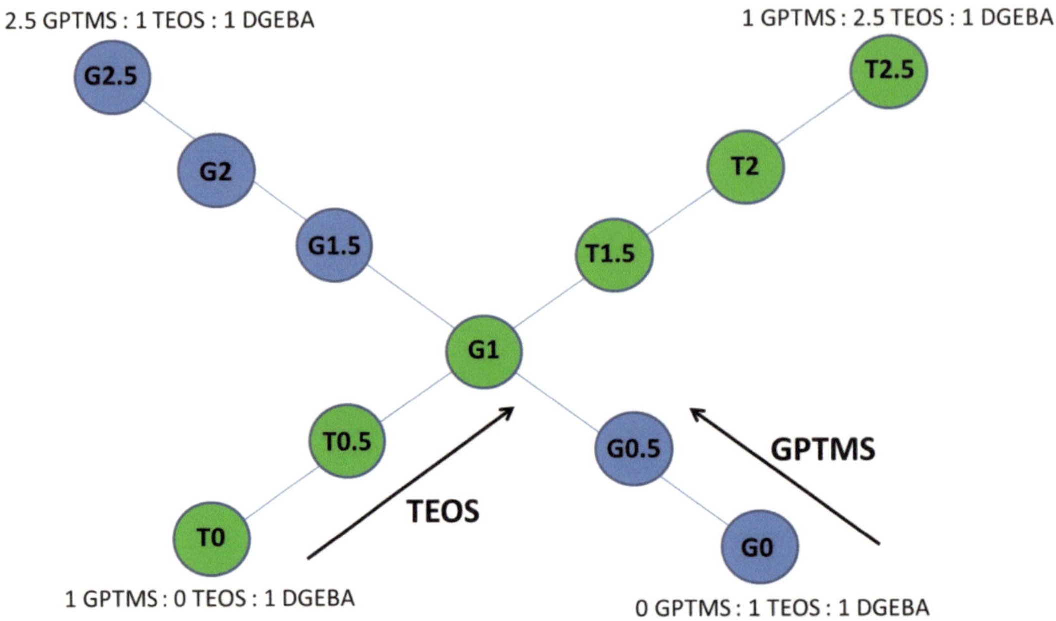

Figure 4. Epoxy-silica hybrids sample names and compositions.

To investigate the relation between structure and barrier properties, the hybrids films were prepared at different synthesis conditions (**Table 1**), varying the ratio between the organic to inorganic phase (MMA/TEOS), the temperature (80–100°C) and time (2–4 h) of the organic precursor reaction, as well as the BPO/MMA molar ratio (0.01–0.1). The molar ratios of H_2O/Si = 3.5 and ethanol/H_2O = 0.5 were kept constant. Ce(IV) salt (ammonium cerium nitrate), lignin, carbon nanotube (CNT), and graphene oxide (GO) were added separately as modifier to the inorganic precursor of the PMMA-silica hybrid.

Carbon steel 1020 (25 mm × 20 mm × 5 mm), a ferrous alloy with low carbon content, and 2024 aluminum alloy (20 mm × 20 mm × 1 mm) have been used as substrates. Low-carbon steels are produced in large quantities at relatively low costs and widely used in automobilist, construction, oil industries, etc [21]. Although the use of ferrous alloys is economically viable due to the low cost and versatility, corrosion is the great obstacle when it comes to the durability of these materials that undergo severe corrosion in contact with humid environments, low amounts of chloride ions and acid solutions in general. The 2000 and 7000 series of aluminum alloys, containing roughly 4.3–4.5% copper, 0.5–0.6% manganese, 1.3–1.5% magnesium, are widely used in the aerospace industry due to their improved mechanical properties; however, they are susceptible to enhanced corrosion especially at the grain boundaries. Prior to deposition, all substrates had been sanded with 100, 300, 600, and 1500 grit emery paper, washed with isopropanol for 10 min in an ultrasound bath and dried under a nitrogen stream. The deposition of the hybrids coatings was performed by dip-coating (Microchemistry—MQCTL2000MP) at a rate of 14 cm min^{-1}, with 1 min of immersion and air-drying during 10 min at room temperature. This procedure was performed three times for each sample. The coated substrates and

Samples	MMA:MPTS:TEOS molar ratio	BPO/MMA molar ratio	Organic phase synthesis	Filler	Reference
M2	2:1:2	0.01	70°C/2 h	–	[22]
M4	4:1:2	0.01	70°C/2 h	–	[22]
M8	8:1:2	0.01	70°C/2 h	–	[22]
M10	10:1:2	0.01	70°C/2 h	–	[22]
M8_4h	8:1:2	0.01	80°C/4 h	–	[18]
M8_4h_E0.2	8:1:2	0.01	80°C/4 h		[10]
M8_T80B0.01	8:1:2	0.01	80°C/4 h	–	–
M8_T90	8:1:2	0.01	90°C/4 h	–	–
M8_T100	8:1:2	0.01	100°C/4 h	–	–
M8_B0.05	8:1:2	0.05	80°C/4 h	–	–
M8_B0.10	8:1:2	0.10	80°C/4 h	–	–
M8_Ce	8:1:2	0.01	70°C/2 h	Ce/Si molar ratio: 0.1, 0.2, 0.3, 0.5, 0.7, 1, 3, 5%	[11]
M8_lignin	8:1:2	0.01	70°C/2 h	lignin: 0.05, 0.10, 0.50, 1.00 wt.%	[20]
M8_CNT	8:1:2	0.01 and 0.05	80°C/4 h	C_{CNT}/Si molar ratio: 0.05%	[18]
M8_GO	8:1:2	0.01 and 0.05	80°C/4 h	C_{GO}/Si molar ratio: 0.05%	[18]

Table 1. PMMA-silica hybrid preparation conditions.

the remaining solution, placed in Teflon holders, were cured for 24 h at 60°C and then 3 h at 160°C to ensure the liberation of all volatile species and the densification of the hybrid matrix.

2.2. Characterization techniques

Structural and morphological characteristics have been investigated using nuclear magnetic resonance spectroscopy (NMR), X-ray photoelectron spectroscopy (XPS), small angle X-ray scattering (SAXS), atomic force microscopy (AFM) and thermogravimetric analysis (TGA). The anticorrosive properties of coated samples were evaluated by exposure of the coated samples to standard 3.5% saline and saline/acid solutions, using electrochemical impedance spectroscopy (EIS).

The thickness of the coatings was determined using a Filmetrics F3-CS optical interference system. An Agilent Technologies Model 5500 atomic force microscope was used to obtain AFM topography images, in tapping mode, with 1 × 1 μm, of the hybrid coatings deposited on the metallic substrates. ^{29}Si nuclear magnetic resonance spectroscopy (^{29}Si-NMR) measurements of the hybrid powders were performed in a 300-MHz Varian Inova spectrometer, using a Larmor frequency of 59.59 Hz and tetramethyl silane (TMS) as an external standard. The

CasaXPS processing software was used for spectral deconvolution using Gauss profiles. XPS was carried out in a UNI-SPECS UHV surface analysis system, using the Mg Kα radiation (hν = 1253.6 eV) and pass energy of 10 eV to record the high-resolution spectra. The near surface composition was determined from relative peak intensities of carbon (C 1s), oxygen (O 1s) and silicon (Si 2p) corrected by Scofield's atomic sensitivity factor of the corresponding elements. To study the oxidation state of Ce (Ce 3d) and the local bonding structure of carbon (C 1s), oxygen (O 1s), and silicon (Si 2p) of the coatings, the spectra were deconvoluted applying Voigt profiles and Shirley's background subtraction using the CasaXPS software. SAXS experiments were carried out at the SAXS-1 beamline in the National Synchrotron Light Laboratory (LNLS, Campinas, Brazil) to determine the nanostructural characteristics of the hybrids. The scattering intensity I(q) was recorded as a function of the modulus of the scattering wave vector $q = (4\pi/\lambda) \sin \theta$, θ being half of the scattering angle. The SAXS beamline uses a monochromatic X-ray beam (λ = 1.548 Å) and a 2D detector, Dectris Pilatus 300k, positioned 0.9 m away from the sample holder. Thermogravimetric analysis of unsupported hybrids films was performed in a TA Instruments STD Q600 analyzer, under a nitrogen flow of 100 mL min^{-1}.

The anticorrosive performance of hybrid coatings, deposited on A1020 carbon steel or Al2024 aluminum alloy, was investigated by electrochemical impedance spectroscopy (EIS) with a Gamry Potentiostat Reference 600, using 10 points per decade and RMS amplitude of 10 mV in a frequency range of 50 mHz–100 kHz. The electrochemical cell consisted of an Ag|AgCl|KClsat reference electrode, a platinum mesh counter electrode, a platinum electrode connected to the reference electrode through a 0.1-μF capacitor and the working electrode of either coated or uncoated metal substrate. The measurements were performed once a week, in saline (3.5% NaCl) or saline/acid solutions (0.05-mol L^{-1} H_2SO_4 + 0.05-mol L^{-1} NaCl), until a significant drop of the impedance modulus was observed, indicating the failure of the coating.

3. Results and discussion

A number of interesting results have been obtained for novel epoxy-silica and PMMA-silica hybrid coatings, concerning their nanostructural properties, modified by the variation of synthesis conditions or by addition of nanofillers, in form of lignin, carbon nanotubes, and graphene oxide. The main purpose of this work was to relate these properties with the barrier characteristics, in terms of corrosion resistance and durability in aggressive environments and to compare the obtained results with those reported for a variety of hybrid coating systems. For the fine tuning of the performance of both coating systems toward an efficient and stable anticorrosive barrier, it is crucial to obtain detailed information on the formation process of the hybrid network and the structural and compositional properties of the nanocomposites.

3.1. Epoxy-silica hybrid

Bisphenol stands for a group of chemical compounds with two hydroxyphenyl functionalities. There is a wide diversity of bisphenol molecules; however, the most common are

http://dx.doi.org/10.5772/67909

the Bisphenol A (BPA) and the Bisphenol F (BPF) (**Figure 5**). Epoxy resins can be produced from the combination of bisphenol, such as bisphenol A, with epichlorohydrin (IUPAC name: 2-(chloromethyl)oxirane)) to give, for example, bisphenol A diglycidyl ether (**Figure 6**). The epoxy resins present in general poor thermal, mechanical, and chemical stability, properties which are however significantly improved when a curing agent is added. Most curing agents are composed of nitrogen-containing molecules that have a functionality equal or superior of three ($f \geq 3$), which provides cross-linking between the bisphenol segments. The functionality is the number of available bonding sites, such as f = 4 for diamino diphenyl methane (4 hydrogens prone to provide bond), f = 6 for triethylene tetraamine, and f = 5 for diethylenetriamine (**Figure 7**). Curing reactions by DETA proceed by S_N2 nucleophilic attack of the curing agent to the less-substituted carbon in the oxirane ring, resulting in its opening and formation of an OH group. The nitrogen of the amine group can attack another epoxy ring resulting in a highly branched polymer system, known as a thermoset, which presents high thermal stability and mechanical resistance [23]. This second nucleophilic attack of nitrogen can occur at the epoxy group of the resin or another molecule containing epoxy group, such

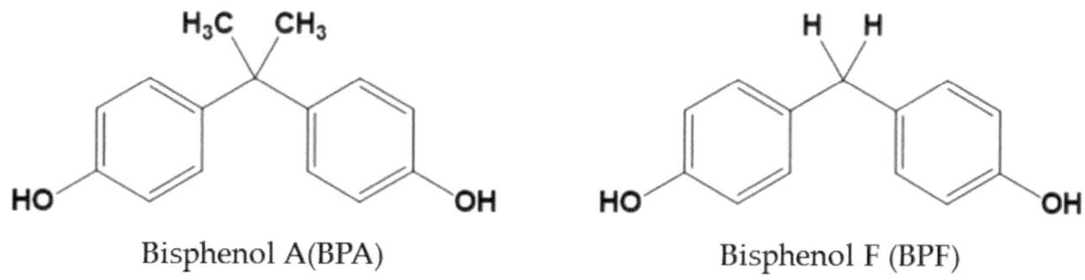

Figure 5. Common epoxy resin precursors.

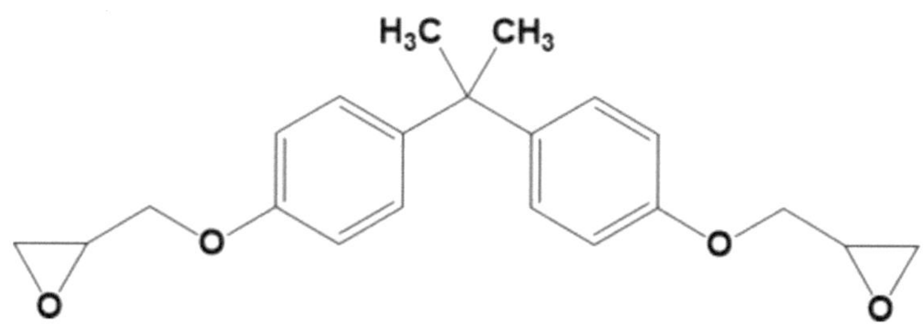

Figure 6. Bisphenol A diglycidyl ether.

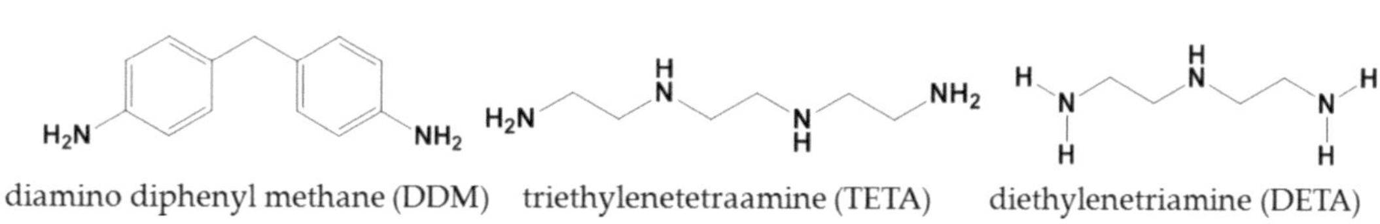

Figure 7. Common curing agents.

as (3-glycidoxypropyl)trimethoxysilane (GPTMS) to produce an organic-inorganic hybrid structure (**Figure 8**).

Simultaneously to the curing reaction, the sol-gel reactions of hydrolysis and condensation take place to produce the silica inorganic phase. GPTMS and TEOS Si–O–R groups, in presence of acidified water, become Si–OH through the hydrolysis reaction, and posteriorly, the Si–OH groups can condense with another Si–OH group or an initial Si–O–R group, forming Si–O–Si bonding and eliminating water or alcohol, respectively.

The surface characterization of the epoxy-silica hybrids deposited on carbon steel has shown that the coatings are uniform, transparent, smooth, and crack free (**Figure 9**). AFM images

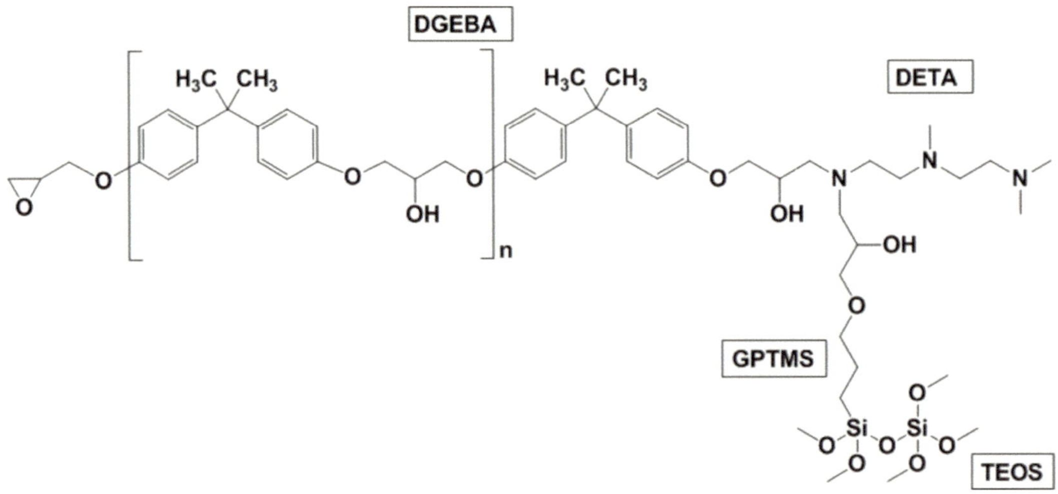

Figure 8. Molecular structure of the epoxy-silica hybrid.

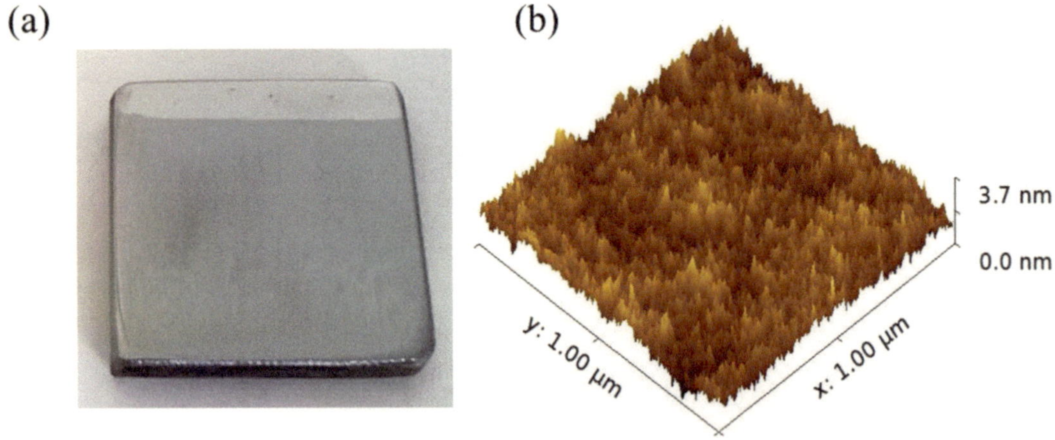

Figure 9. Representative image (a), and AFM image (b), of T1/G1 epoxy-silica hybrid coating deposited on carbon steel.

with an area of 1 μm² were used to obtain the surface roughness of the coatings. **Table 2** summarized all RMS surface roughness and thickness values determined for epoxy-silica hybrids of T-series (TEOS variation) and G-series (GPTMS variation). With increasing GPTMS and TEOS fraction, a significant increment of the surface roughness can be observed. The data suggest that increasing concentration of TEOS has a larger impact on the surface roughness than that of GPTMS, probably due to the formation of silica domains of larger size. Measurements of the films thickness indicate for all samples of the G-series a constant value of about 1.7 μm, while for films of the T-series the thickness varies from 2 to 3 μm, except for the T1.5 sample having 6.7 μm.

Samples	Thickness (μm)	RMS roughness (nm)	C_d (%)	α	R_g (nm)	d (nm)	T_0 (°C)	EIS lifetime (days)
T0	3.0	0.3	–	–	–	3.7	–	–
T0.5	2.0	0.7	85.0	3.8	–	4.2	306	2
T1/G1	1.6	1.2	83.8	3.2	0.8	–	293	5
T1.5	6.7	0.3	87.1	3.9	0.8	–	314	42
T2	3.3	2.6	87.8	3.5	0.8	–	295	1
T2.5	1.8	5	–	3.4	0.8	–	302	4
G0.0	1.0	–	–	3.8	–	–	–	–
G0.5	1.8	0.6	87.8	1.8	1.5	–	285	55
G1.5	1.8	0.4	94.7	4.0	0.6	–	297	–
G2	1.6	1.0	–	4.0	0.4	–	304	1
G2.5	1.8	1.4	–	4.0	0.3	–	306	2

Table 2. Properties of epoxy-silica hybrids: film thickness (optical interferometry); surface roughness (AFM); degree of polycondensation, C_d, (^{29}Si-NMR); Porod coefficient, α, radius of gyration, R_g, and correlation distance, d, (SAXS); temperature of the limit of thermal stability T_0 in N_2 atmosphere (TGA); and coating lifetime in 3.5% NaCl (EIS).

	G0.5		G2.5		T0.5		T2.5	
	XPS	Nominal	XPS	Nominal (at.%)*	XPS	Nominal	XPS	Nominal
Si 2p	4.8	4.3	6.3	6.0	4.4	4.0	7.5	7.6
C 1s	67.7	71.0	62.7	66.0	70.1	71	62.1	61.8
O 1s	25.5	22.0	28.1	25.0	22.9	22.3	28.1	28.1
N 1s	2.0	2.5	2.8	3.5	2.5	3	2.4	2.6
C/Si	14.1	16.9	10.0	11.0	15.9	19	8.3	8.1

*XPS experimental error ±5%.

Table 3. Comparison between XPS and calculated nominal atomic concentrations for epoxy-silica coatings.

Table 3 shows that results of the quantitative XPS analysis are in good agreement with those obtained for the nominal composition for both series of samples. As expected, the data show an increase of silicon and oxygen atomic concentration for the G and T-series, while nitrogen content increases slightly only for the G-series due to the higher DETA content. As the structure of GPTMS contains also carbon atoms, its addition leads to a less pronounced increase of the Si content. As a consequence, for the G-series the decrease of the C/Si ratio from 14.1 (G0.5) to 10.0 (G2.5) was smaller than that observed for the T-series from 15.9 (T0.5) to 8.3 (T2.5).

The chemical bonding structure of the inorganic network can be characterized according to the proportion of different Si species having a fixed number of oxygen bridging silicon atoms bonded to one (central) silicon atom. A common notation is Q^j for orthosilicates ($0 \leq j \leq 4$), such as TEOS, and T^j for organically modified silicates ($0 \leq j \leq 3$), such as GPTMS, where j gives the number of Si–O–Si bridges attached to the silicon atom.

Figure 10 shows the ^{29}Si NMR spectra, fitted with Gaussian components, used to extract the proportion of Q^j and T^j species. It can be observed that the Q^4 and T^3 peaks (−107 ppm and −62 ppm, respectively) have the highest intensities in relation to the other components related to lower network connectivity. The degree of connectivity of the inorganic phase, the so-called degree of polycondensation, C_d, has been determined from the fitted ^{29}Si NMR spectra using the following equation:

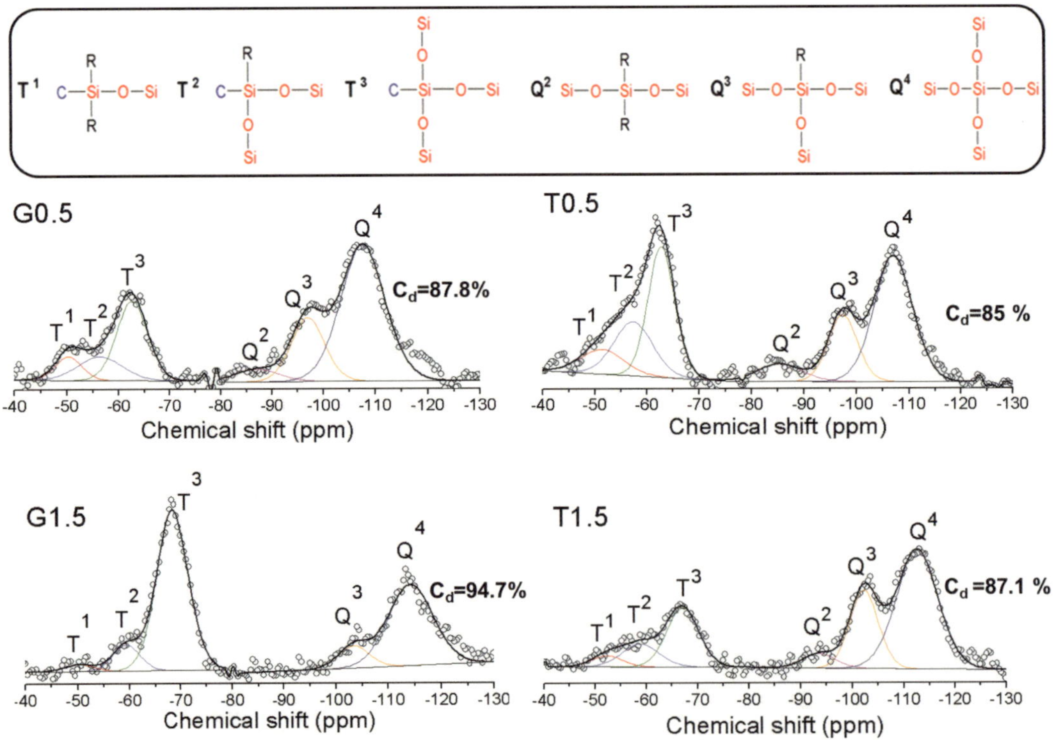

Figure 10. ^{29}Si-NMR spectra obtained for epoxy-silica hybrids. Inset: schematic representation of the T^J and Q^J species, where 'R' indicates OH or OCH_3 or OCH_2CH_3 groups.

http://dx.doi.org/10.5772/67909

$$C_d = \frac{T^1 + 2T^2 + 3T^3}{3} + \frac{Q^1 + 2Q^2 + 3Q^3 + 4Q^4}{4} \times 100 \quad (4)$$

The C_d values of **Table 3** show a high connectivity of the inorganic network with a clear predominance of a tetra-substituted TEOS and a tri-substituted GPTMS sites. Furthermore, it seems that an increase of GPTMS favors the Q^4 and T^3 structures, yielding a highly cross-linked inorganic network reaching about 95% connectivity for G1.5 sample, while the variation of TEOS does not change the C_d values significantly, remaining in the range of 85–88%. More information on the structure and size of the inorganic domains was obtained by small angle X-ray scattering (SAXS) measurements.

The SAXS technique allows to access the nanostructural characteristics of the inorganic network due to the higher electronic density of silica compared that of the polymeric matrix. The log-log plots of scattering intensities I (q) recorded for different fractions of GPTMS and TEOS (**Figure 11**) show three main characteristics: a linear decay located at low q values, corresponding to the Porod region; a Gaussian decay in the mid q-range, corresponding to the Guinier regime; and a broad correlation peak superimposed to the Guinier region, observed only for T0.0 and T0.5 samples. The former feature, in the mid q-range, is characteristic of a diluted set scatters, while the latter is the result of the interferences of the scattered X-ray caused by the concentrated set of nano-objects.

These scattering patterns have been already observed for other silica-polymer hybrids [24, 25] and attributed to a hierarchical organization of silica nano-domains. Accordingly, we propose that the nanostructure of the hybrid can be described by a two-level hierarchical model, corresponding to a diluted or concentrated (T0.0 and T0.5) set of silica nanoparticles inside the aggregation zones embedded in the polymer matrix. In the case of the diluted system, the size of the smaller particles was determined, in terms of the radius of the gyration, R_g, by fitting the Gaussian decay observed in the mid q-range using the

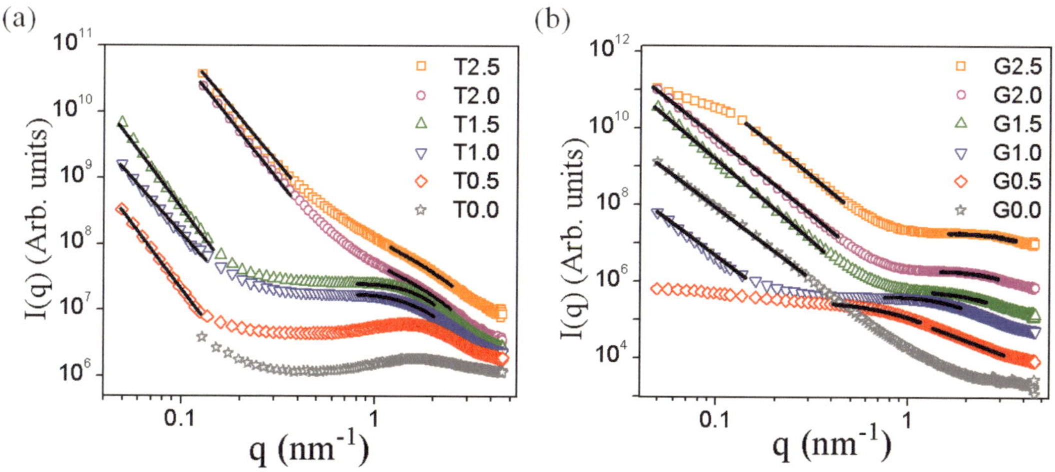

Figure 11. SAXS curves of T-series (a), and G-series (b), where the black lines represents the fits used to calculate the Porod coefficient (α) for q < 1 (except for the G0.5 sample, which presents α at q>1), and the radius of the gyration (R_g) for q >1. (The intensities were shifted to obtain a better visualization of the curves).

Guinier model: $I(q) = I_0 \exp(-R_g q^2/3)$, where I_0 is the scaling factor. However, this was only possible for scattering curves, which did not present an overlapping correlation peak. Therefore, values for R_g and those for the correlation distance, $d \approx 2\pi/q$, have been obtained only for a restricted number of samples (**Table 2**). Except for T0.0, the form of the scattering objects was determined by fitting the curves using the Porod model: $I \propto q^{-\alpha}$, where α is the Porod exponent. $\alpha \approx 4$ indicates a bi-phase system formed by set of nearly isometric scattering objects with a smooth surface, while for smaller values, a rough surface (fractal) is expected.

The results indicate that the inorganic phase consists of aggregates with relatively smooth surface and an average spacing of several nanometers ($d \approx 4$ nm). These domains have been formed by agglomeration of smaller silica particles with a size of about 1 nm ($0.3 < R_g < 1.5$ nm).

Some clear correlations between these parameters and the increasing silica concentration of the G- and T-series could be established. For the G-series, the evolution of the SAXS pattern evidences the role of GPTMS in controlling the size of primary silica particles and they aggregation. The power law decay over a decade and $\alpha = 3.8$ observed for the hybrid prepared only with TEOS (G0.0) characterizes the scattering by the surface of very large silica particles (>30 nm). The addition of a small amount of GPTMS (G0.5 sample) reduces the size of silica particles more than ten times ($R_g = 1.5$ nm) and prevents further aggregation, as evidenced by extended plateau at $q < 0.1$ nm. These unique features suggest for G0.5 sample an elevated nanostructural homogeneity, which might be responsible for the superior corrosion protection performance of this material (**Table 2**). In the case of the T-series, the correlation peak disappears for higher TEOS content and the linear decay shifts to higher q-values. These features evidence that TEOS addition favors the formation of more open aggregates, leading to a less compact nanostructure.

The thermal properties of the hybrids were studied by thermogravimetry under nitrogen flow. **Table 2** shows the temperature of the limit of thermal stability, T_0, for all epoxy-silica hybrids, defined as the temperature of 5% weight loss during the annealing process. The hybrids presented a thermal stability of about 300°C, relatively high values compared with those of other polymeric and hybrid materials [18, 22]. This advantageous property comes from the highly cross-linked structure provided by the curing agent (DETA) combined with the high polycondensation degree of the silica phase, as revealed by ^{29}Si NMR.

The anticorrosive performance of the hybrids was assessed by EIS measurements, in a 3.5% NaCl saline solution at 25°C. The hybrid coatings deposited on carbon steel were attached to an electrochemical cell, and after verifying a constant value of the open-circuit potential, the impedance measurements were performed as a function of time until a significative drop of the impedance modulus occurred. This time period was defined as lifetime of the coating, listed in **Table 2**. The impedance modulus at low frequency of the Bode plot is generally used as an indicator of the anticorrosive performance of the coating, with values higher than 0.1 GΩ cm^{-2} typically considered an excellent protection. The corrosion resistance of the films generally decreases with time, caused by the penetration of electrolyte into the protective layer through zones of residual porosity and defects.

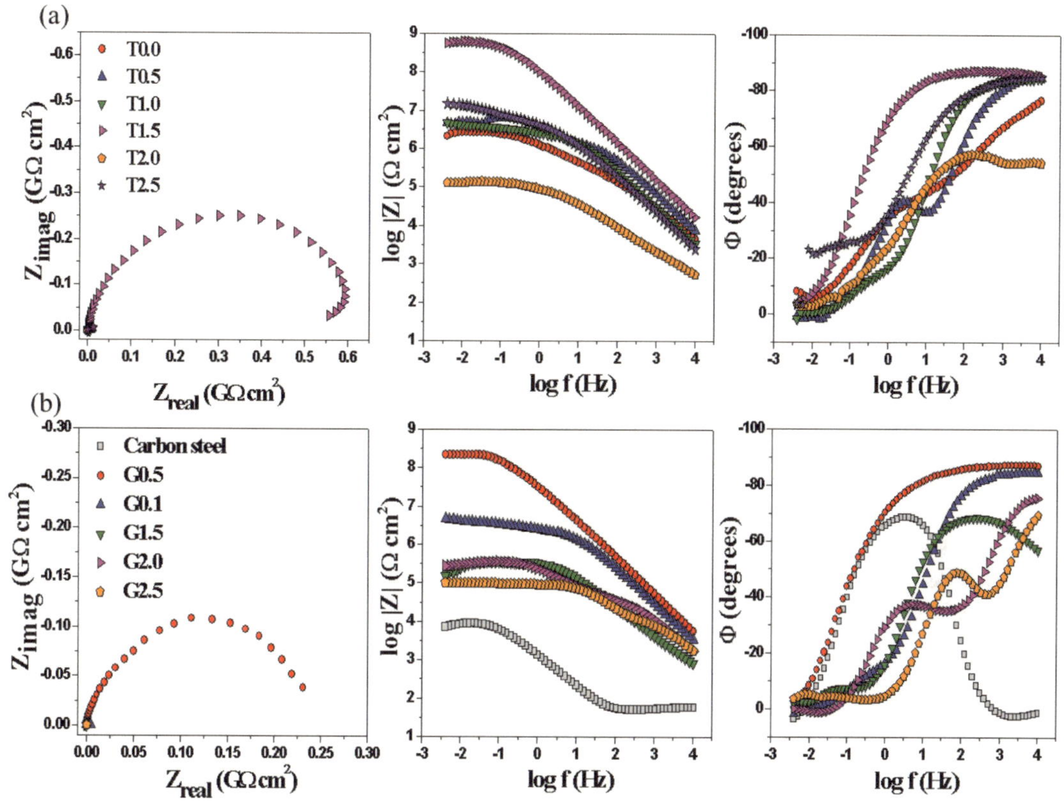

Figure 12. Nyquist and Bode plots of (a) series T and (b) series G of the epoxy-silica coatings deposited on carbon steel, compared to those of bare carbon steel, after 1 h of immersion in 3.5% NaCl solution.

The Nyquist and Bode plots obtained after 1 day of immersion in 3.5% NaCl solution are presented in **Figure 12**. It can be observed that two samples containing intermediate TEOS to GPTMS ratios (T1.5 and G0.5) presented the highest impedance modulus of 0.9 and 0.2 GΩ cm^2, respectively, and showed also the longest lifetime of several weeks (**Table 2**). These coatings show at higher frequencies (>1 Hz), a capacitive behavior with a phase angle higher than −80° extending over a range of 4 decades, characteristic for an efficient anticorrosive barrier layer. In contrast, for formulations with excess of TEOS or GPTMS, both the corrosion resistance and lifetime values show considerably lower values.

The results of the structural analysis indicate that the excellent barrier properties, found for coatings with intermediate TEOS to GPTMS ratio, result from a highly reticulated hybrid structure combining a number of favorable properties, such as a high polycondensation degree of the inorganic phase, a extremely smooth surface, indicating a very homogeneous distribution of silica nanodomains, high thermal stability, as well as an adequate quantity of the silica phase which ensures a good adhesion of the film to the metallic substrate. Although the corrosion protection efficiency of the best epoxy-silica coatings, reported so far [7, 26, 27], is comparable with results presented in this work, it may profit from their 10–100 times higher thickness.

3.2. PMMA-silica hybrids

Poly(methyl methacrylate), also known as acrylic and Pexiglas®, is a rigid, low cost, nontoxic, transparent and colorless thermoplastic polymer, extensively used as optical lenses, protective coatings, optical fibers, and as an alternative to glass in windows as well as a variety of household appliances. The introduction of an inorganic component, such as silica, improves the thermal stability, mechanical strength, and the adhesion to metallic substrates, the latter property being an essential feature for a high-efficiency coatings. The covalent bond between the PMMA and the silica phase can be achieved by the addition of 3-(trimethoxysilyl)propyl methacrylate (MPTS), a coupling agent formed by an alkoxy-silane group attached by a non-hydrolysable Si–C bond to the acrylic tail, which polymerizes with PMMA chains, while the inorganic part reacts with the silica precursor (TEOS), yielding an organic-inorganic hybrid structure, shown in **Figure 13**.

A variety of PMMA-silica hybrids have been studied, changing the organic/inorganic phase proportion, the amount of thermal initiator, the synthesis temperature and time, as well as the ethanol-to-water ratio. Furthermore, cerium salt has been added to the PMMA-silica matrix as corrosion inhibitor, and lignin, carbon nanotubes, and graphene oxide as fillers. The main results, found for pure PMMA-silica hybrids, are summarized in **Table 4**, those obtained using additives will be discussed in the following sections.

Similar to epoxy-silica coatings, PMMA-silica hybrids deposited on metallic substrates were homogeneous, transparent and had a very smooth surface. Structural analysis of PMMA-silica hybrids, performed by AFM, SAXS, NMR and XPS, has shown that the nanostructure is formed by a dense amorphous network of ramified silica-siloxane cross-link nodes, covalently interconnected by short PMMA chain segments [10, 22]. Varying the MMA/MPTS molar ratio from 2 to 10, NMR and SAXS results have shown that the M8 sample (MMA/MPTS = 8) presented the highest degree of polycondensation (83.9%) of the silica nanoparticles with an average radius of 0.8 nm, spaced by PMMA segments over an average distance of 4.6 nm. This coating exhibited also an excellent adhesion to the substrate (detachment force > 3.5 MPa) and the best anti-corrosion performance [10, 22]. EIS and potentiodynamic polarization results have shown that the M8 coating deposited on carbon steel acts as a very efficient corrosion barrier, increasing the total corrosion resistance by almost 6 orders of magnitude (>1 $G\Omega\ cm^2$) and reducing the current densities by more than 4 orders of magnitude (<0.1 $nA\ cm^{-2}$), compared to the bare steel substrate [22]. Furthermore, XPS analysis confirmed that no corrosion-induced changes had occurred after 18 days of immersion in 3.5% NaCl solution [22].

Increasing the synthesis temperature from 70 to 80°C and the time of reaction from 2 to 4 h (sample M8_4h), an increase in the amount of polymeric phase was detected yielding a more compact and durable coating (56 days) [18, 22]. After optimizing also the ethanol-to-water ratio of the inorganic phase to a value of 0.2, the corrosion resistance and lifetime were further increase to 196 days in 3.5% NaCl (M8_4h_E0.2 coating, **Table 4**) [10]. Other important finding was the improvement of the corrosion resistance by hot deposition (M8_T80B0.01), which enhances the reaction between the inorganic phase and the metal substrate, improving the coating adhesion. This sample has been also deposited on Al2024 substrate and tested in saline (**Figure 14a**) and saline/acid environment (**Figure 14b**). This coating highlights a

http://dx.doi.org/10.5772/67909

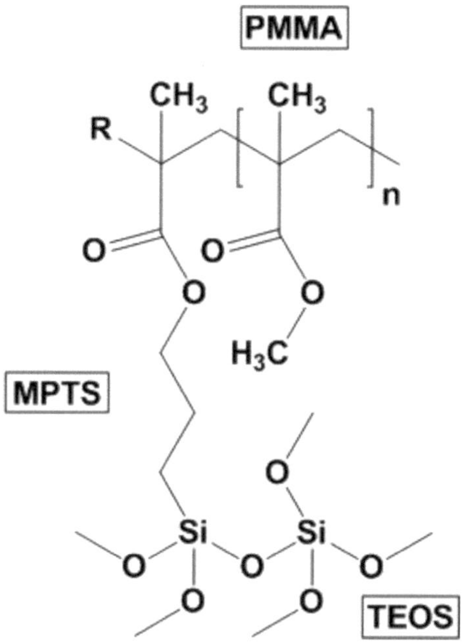

Figure 13. Molecular structure of the PMMA-silica hybrid.

Samples	Thickness (μm)	RMS roughness (nm)	C_d (%)	T_0 (°C)	|Z| (GΩ cm²)	EIS lifetime (days)	Reference
M2	1.5	–	80.9	–	~0.001	–	[21]
M4	–	–	79.7	–	~0.01	–	[21]
M8	3	0.3	83.9	–	~1	18	[21]
M10	–	–	75.8	–	~0.001	–	[21]
M8_4h	2.8	0.4	78.0	205	~0.1	56	[18]
M8_4h_E0.2	2.0	0.5	82.0	230	~3	196	[10]
M8_T80B0.01	2.8	–	–	230	~3	40	–
M8_T80B0.01*	3.0	–	–	–	~50	560	–
M8_T90	2.6	–	–	238	~5	34	–
M8_T100	2.5	–	–	250	~0.1	35	–
M8_B0.05	5.0	–	–	270	~10	42	–
M8_B0.10	9.7	–	–	223	~10	>186	–

*Al2024 substrate.

Table 4. Properties of PMMA-silica hybrid coatings: film thickness (optical interferometry); surface roughness (AFM); degree of polycondensation, C_d, (^{29}Si-NMR); limit of thermal stability T_0 in N_2 atmosphere (TGA); impedance modulus |Z|, after 1 day exposure to 3.5% NaCl solution (EIS); and coating lifetime in 3.5% NaCl (EIS).

corrosion resistance of up to 50 GΩ cm², in saline environment, showing only a small performance decrease to 0.1 GΩ cm² after 560 days exposure. This is to our best knowledge the highest durability, obtained so far for hybrid coatings in standard saline solution. Also in contact with a saline/acid solution (0.05 mol L^{-1} H_2SO_4 + 0.05 mol L^{-1} NaCl), this about 3-μm thick coating presented a high corrosion resistance (20 GΩ cm²), remaining almost unchanged during its lifetime of 87 days. It is interesting to note that the phase angle dependence has a capacitive behavior ($\theta \approx -90°$), over a frequency range of 6 decades, a behavior close to that of an ideal capacitor, highlighting the extraordinary performance of this coating.

PMMA-silica hybrids have been also prepared at different synthesis temperatures of the organic precursor (80–100°C) and different BPO/MMA molar ratio (0.01–0.1), using the well-established MMA/MPTS molar ratio of 8 [22]. The increase in the synthesis temperature did not influence significantly the structure, the thermal properties and the corrosion resistance, however, the increase of the BPO amount led to an increase of the polymerization degree, thermal stability of 40°C (BPO0.05), and also of the anticorrosive efficiency (**Table 4**). The sample M8_BPO0.05 and M8_BPO0.10 presented an impedance modulus of 10 GΩ cm² in a saline medium (3.5% NaCl), remaining essentially unchanged during more than 6 months of exposure (M8_BPO0.10).

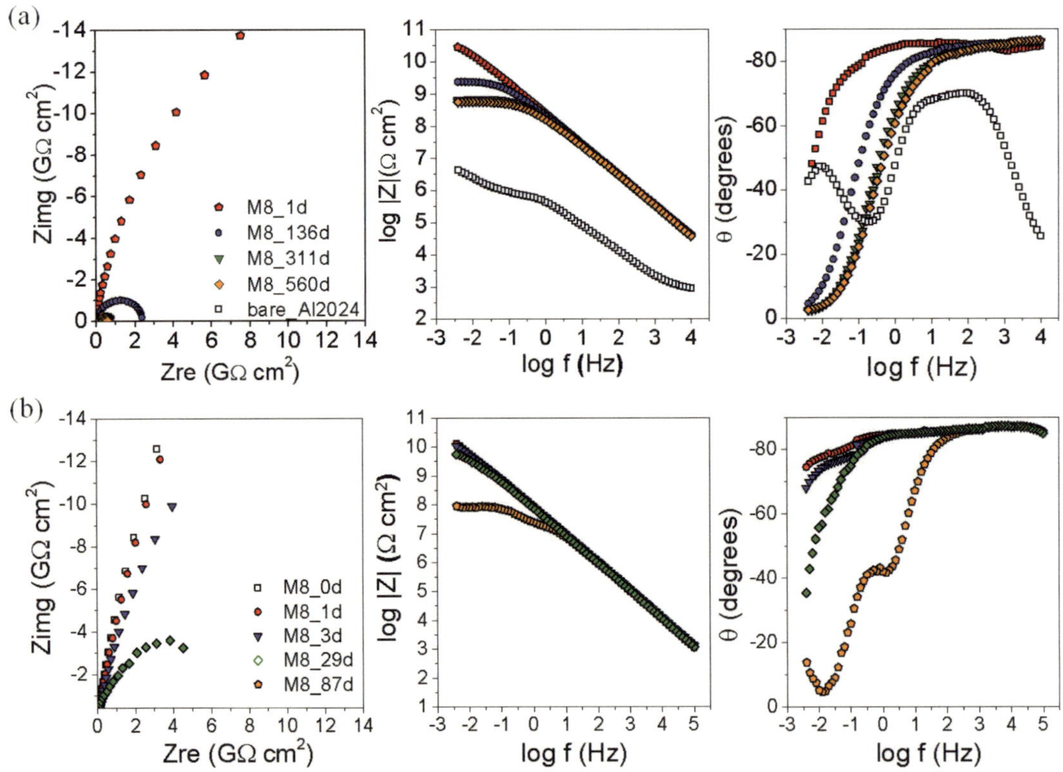

Figure 14. Time dependence of Nyquist and Bode plots of the M8_T80BPO0.01 PMMA-silica coating deposited on Al2024 substrate in contact with a) in 3.5% NaCl solution and b) 0.05 mol L^{-1} NaCl + 0.05 mol L^{-1} H_2SO_4 solution.

3.3. PMMA-silica hybrid modified with Ce(IV) salt for self-healing ability

After identifying the optimum proportion of polymeric and silica phase for the formation of a highly ramified structure (MMA/MPTS/TEOS molar ratio = 8/1/2), increasing molar percentage of Ce(IV) ions (0.1% < Ce/Si < 5%) have been added to the inorganic precursor to enhance the passivating character of the films [11].

NMR, XPS, and SAXS results, summarized in **Table 5**, have revealed the active role of Ce(IV) in the PMMA-silica matrix. The correlation of XPS and NMR data evidenced that the Ce(IV) concentration is directly related to the polycondensation degree (C_d) and the degree of Ce(IV) reduction, both decreasing with increasing cerium concentration. Low concentrations of cerium lead to an enhanced polycondensation of the siloxane/silica phase, with connectivity of the inorganic phase up to 87%. For low doping levels of Ce/Si < 0.7%, SAXS results have revealed increasing values of radius of gyration, R_g, suggesting an active role of Ce(IV) as oxidation agent in the enhanced growth of a cross-linked and polycondensed inorganic phase. Detailed investigation of the structural effects of cerium species has shown that reduction of Ce(IV) ions not only catalyzes a higher connectivity of the silica phase, but also enhances the polymerization of organic moieties. The resulting enhancement of the overall network connectivity leads to an improvement of the thermal stability of the hybrids, as evidenced by the results of the thermogravimetric analysis [11].

The electrochemical assays, performed by EIS and potentiodynamic polarization curves, have shown that the PMMA-silica coatings containing intermediate concentrations of cerium present a combination of high-corrosion resistance (~10 GΩ cm^2), elevated overpotential stability at low-current densities (<10^{-11} A), as well as excellent long-term stability of up to 304 days. Compared to the bare carbon steel substrate, the coated samples showed up to 6 orders of magnitude higher impedance modulus and up to 6 orders of magnitude lower current densities [11].

For coatings containing elevated Ce(IV) doping levels (Si/Ce = 5%), the self-healing effect was observed, induced by the formation of insoluble cerium oxides and hydroxides in corrosion affected regions. The presence of these phases in the near surface region was evidenced by XPS O 1s spectra and by scanning electron microscopy, showing the presence of nanopits (<300 nm). It was suggested that these phases were formed by reactions of Ce(III) and Ce(IV) with water and residual hydroxyl groups of the hybrid. The self-healing process prevented the progression of the corrosion process for more than 13 months keeping the corrosion resistance constant above 0.01 GΩ cm^2. The excellent anticorrosive efficiency achieved by PMMA-silica coatings containing cerium can be related to a double effect of Ce(IV), combining the densification of the hybrid network with the formation of insoluble cerium species in regions affected by pitting [11].

3.4. PMMA-silica hybrid reinforced with lignin, carbon nanotubes and graphene oxide

Lignin is a macromolecule present in the cell walls of terrestrial plants that confers rigidity and impermeability, usually corresponding to 15–30% of the dry weight of wood (**Figure 15a**).

Properties	Ce(IV)/Si molar fraction (%)								
	0	0.1	0.2	0.3	0.5	0.7	1	3	5
C_d (%)	82.8	87.1	–	83.6	82.7	79.3	78.5	77.6	77.3
XPS Ce(IV) fraction (%)	–	–	–	28.5 ± 4	37.2 ± 3	46.4 ± 2.5	48.5 ± 2.5	55.5 ± 2	60.4 ± 1.5
R_g (nm)	0.9	1.1	1.8	2.3	2.3	1.9	1.9	–	–
|Z| (GΩ cm²)	~0.5	~0.5	–	–	~0.1	~10	~5	–	~0.5
Lifetime (days)	42	85	–	96	–	304	65	48	404

Table 5. Properties of PMMA-silica hybrids containing different amounts of Ce(IV): degree of polycondensation, C_d, (^{29}Si-NMR); percentage of the Ce(IV) oxidation state (XPS Ce 3d spectra); radius of gyration, R_g, (SAXS); and impedance modulus, |Z|, after 1 day exposure to 3.5% NaCl solution (EIS).

Presently, millions of tons of lignin are generated from biodiesel and ethanol production, and most part is incinerated to generate electric energy. However, nobler applications have been found to add value to this biomass, such as the reinforcement of different classes of materials. Properties, such as low density, low abrasive character, hydrophobicity, and low cost, make lignin ideal to use as filler in polymeric and organic-inorganic hybrid matrices [20].

Carbon nanotubes (CNTs) and graphene oxide (GO) (**Figure 15b and c**) are also very interesting nanofillers for the structural reinforcement of polymeric and hybrid materials, due to their exceptional thermal, chemical, and mechanical resistance. Both present a hexagonal sp^2 arrangement of carbon atoms, forming extremely stable cylindrical and monolayer structures, respectively. Graphene oxide has been obtained from oxidation and exfoliation of graphite, yielding a graphene layer containing oxygen functional groups such as epoxy, hydroxyl, and carboxyl [18].

PMMA-silica hybrids reinforced with 0.05, 0.10, 0.50 and 1.00 wt.% of lignin have been deposited on carbon steel by dip-coating, producing about 2.5 μm thick coatings (**Table 6**). Optical microscopy and optical microscopy and atomic force microscopy showed that lignin was well dispersed in the hybrid matrix, and all coatings presented a low RMS surface roughness between 0.3 and 0.4 nm. The introduction of lignin in the PMMA-silica hybrid increased the water contact angle of the film surface from 79.3° to 87.9°, the hardness from 22.9 to 30.9 HV, and the scratch resistance (critical load for delamination) from 55 to 80 mN. In addition, the thermal degradation events, obtained by thermogravimetric analysis (TGA) under nitrogen atmosphere, were shifted to higher temperatures with lignin addition, due to its phenolic structure which has the ability to trap radicals formed during the depolymerisation. Besides increasing the thermal stability of the polymeric phase, it acts also as UV stabilizer [20].

The electrochemical tests performed by EIS showed that the PMMA-silica coatings containing lignin act as efficient diffusion barriers, with corrosion resistance higher than 0.1 GΩ cm² after

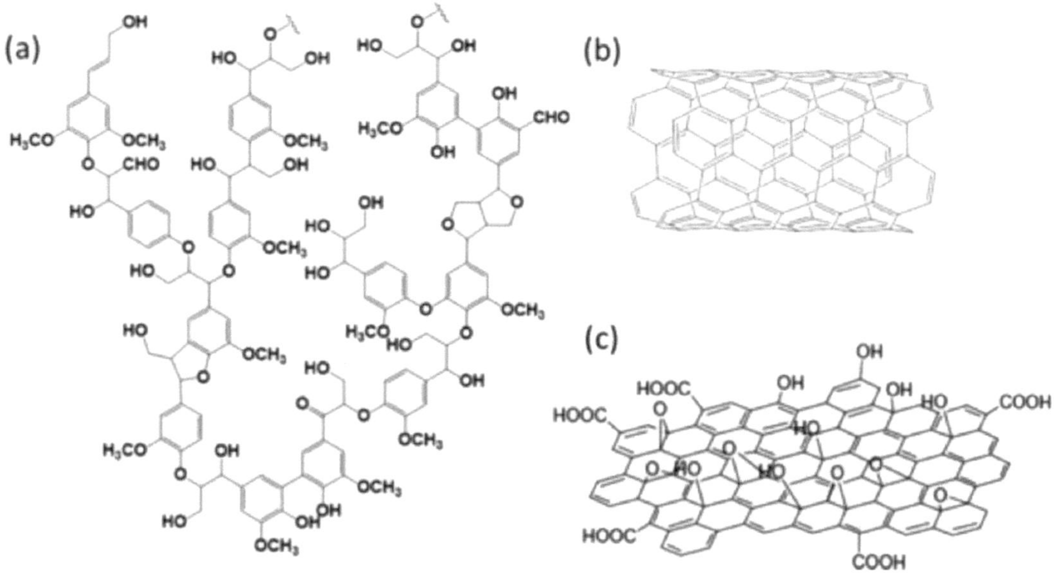

Figure 15. Molecular structure of (a) lignin, (b) carbon nanotube, and (c) graphene oxide.

Sample	Thickness (μm)	T_0 (°C)	Critical load (mN)	|Z| (GΩ cm²)	EIS lifetime (days)	Reference
M8 (70°C/2 h)	2.4	170	55	~0.1	18	[20]
M8_lignin_0.50 wt.%	2.6	180	65	~0.5	–	[20]
M8_lignin_0.10 wt.%	2.5	200	80	~0.5	50	[20]
M8_lignin_0.50 wt.%	2.7	190	66	~0.01	–	[20]
M8_lignin_1.00 wt.%	2.7	195	56	~0.005	–	[20]
M8_BPO001 (80°C/4 h)	2.8	205	78	~0.5	56	[18]
M8_BPO001_ CNT_0.05%	5.7	220	–	~0.1	43	[18]
M8_BPO001_ GO_0.05%	3.1	275	94	~1	203	[18]
M8_BPO005 (80°C/4 h)	3.5	208	84	~1	21	[18]
M8_BPO005_ CNT_0.05%	6.6	209	133	~10	7	[18]
M8_BPO005_ GO_0.05%	5.5	216	148	~5	168	[18]

Table 6. Properties of PMMA-silica hybrids containing lignin, CNT or GO: film thickness (optical interferometry); limit of thermal stability T_0 in N_2 atmosphere (TGA); critical load for delamination (microscrach test); impedance modulus |Z|, after 1 day exposure to 3.5% NaCl solution (EIS); and coating lifetime in 3.5% NaCl (EIS).

exposure to 3.5% NaCl aqueous solution. For intermediate lignin content of 0.10 wt.% the coatings presented best results with an impedance modulus of 0.5 GΩ cm^2, remaining almost unchanged after 50 days of exposure to aggressive environment [20].

Recent studies on the inclusion of carbon nanotubes and graphene oxide in hybrid and polymer matrices have shown excellent results in terms of increased mechanical strength, scratch and wear resistance, thermal stability, adhesion to metallic substrate, hydrophobicity, and electrical conductivity [18, 19, 28–30]. Despite all of these advances, a simultaneous improvement not only of mechanical and thermal stability but also of anticorrosive efficiency of protective coatings has been accomplished only recently by the incorporation CNT and GO in a PMMA-silica matrix [18].

To synthesize the CNT and GO modified PMMA-silica hybrids, first, the single-walled carbon nanotubes and graphene oxide were dispersed in a water/ethanol, adding in the case of CNTs dodecyl sulfate surfactant (SDS) as dispersant. Then, the carbon nanostructures were added to the inorganic precursor solution of the PMMA-silica hybrid at a carbon (CNT or GO) to silicon (TEOS and MPTS) molar ratio of 0.05% in two different matrices, prepared at BPO/MMA molar ratios of 0.01 and 0.05. As the function of BPO as a thermal initiator is to produce radicals that initiate the polymerization process of MMA, an increased BPO content results in enhanced polymerization degree in the hybrid. The transparent hybrids deposited on A1020 carbon steel substrates by dip coating presented thickness values between 2.8 and 6.6 μm (**Table 6**), a good dispersion of the carbon nanostructures, and a very smooth surface (0.3–0.5 nm RMS surface roughness) [18].

Microscratch and wear tests, performed with a spherical-conical diamond tip of 10 μm radius, confirmed for the PMMA-silica coatings that both additives, CNT and GO, improved the scratch resistance (increase of the friction coefficient by 0.1–0.2), adhesion to the metallic substrate (no delamination for M8_BPO001_CNT up to 240 mN) and wear resistance (smooth and shallow wear track after 50 cycles). The superior behavior of CNT containing coatings was attributed to their property to act as rigid obstacles for the scratch tip. Results of the thermogravimetric analysis have shown that the addition of CNT and GO to the BPO0.01 matrix and to a smaller extent to the BPO0.05 matrix, increased the thermal stability of the hybrids up to 70°C for GO containing samples (**Table 6**). This improvement was attributed to interaction between carbon nanostructures and macroradicals generated during the process of depolymerisation combined with a 2D barrier effect of GO, hindering molecular diffusion through the matrix and thus providing an improved thermal resistance [18].

Results of electrochemical impedance spectroscopy in 3.5% NaCl solution showed that PMMA-silica coatings reinforced with CNTs and GO had an improved anticorrosive efficiency, with impedance modulus of ~1 GΩ cm^2 and ~10 GΩ cm^2 for the BPO0.01 and BPO0.05 matrix, respectively. Besides the improved barrier property, GO containing coatings presented also a prolonged lifetime of up to 203 days. This was attributed to the two dimensionality of the GO structure that provides an enhanced barrier effect against the propagation of corrosive species. Furthermore, it was suggested that both carbon nanostructures act as densifiers of the nanocomposite and also as negatively charged repulsive agents for chloride anions, thus improving barrier property of the coating. Based on the equivalent circuit used to fit the EIS data, this notable barrier behavior was interpreted, in terms of two distinct dielectric layers, one related to a porous water uptake zone at the coating/electrolyte interface and the other corresponding to the underlying unaffected film region, having three orders of magnitude higher resistivity [18].

3.5. Advances in organic-inorganic hybrid coatings for corrosion protection

To be able to evaluate the relevance of the obtained results, it is important to place them in the context of the state of the art in the field of anticorrosive coatings. In the last decade, the concept of organic-inorganic hybrids as protective coating has been intensely investigated using different approaches involving a variety of organic and inorganic precursor reagents, resulting in a number of promising coatings systems. The most widely applied formulations for hybrid phases used to prepare high-performance anticorrosive coatings are based on epoxy-silica (**Table 7**) and acrylic-silica (**Table 8**) hybrids, and to a lower extent on polyurethane-silica, polyurethane-silica-zirconia and other epoxy systems (**Table 9**). As can be inferred from these data, the electrochemical barrier properties, obtained for different hybrid formulations, have achieved a notable performance in the last years, making these novel nanocomposites very promising candidates for efficient corrosion protection of metallic surfaces. This is justified especially when considering that a high-corrosion resistance and long durability in aggressive environments can be achieved by thin films with thicknesses of less than 10 μm, resulting in substantially reduced material costs compared to conventional high-performance coating systems. More specifically, regarding the epoxy-silica hybrid system, the results presented in this work and those listed in **Table 7** show that different compositions applied to distinct alloys can provide a very effective long-term corrosion protection [7, 26, 27]. Very promising results were also achieved for PMMA-silica coatings [8, 10, 31], with the highest observed durability of more than 560 days in 3.5% NaCl, and for hybrids containing reinforcement and inhibitor agents [11, 18]. Moreover, for some polyurethane-silica and polyurethane-zirconia-silica systems, it has been demonstrated that they also have a high potential to be used as efficient anticorrosive barrier layers [9, 32].

All these results demonstrate the wealth of possibilities to prepare nanocomposite materials based on organic-inorganic hybrids in the form of highly efficient anticorrosive coatings. The optimization of the barrier property can be achieved by the careful adjustment of the precursor proportions, including coupling agents and additives, together with the conditions of synthesis, deposition, and thermal treatment. However, the main key for this task is an in-depth knowledge of the formation mechanisms as well as the compositional and structural properties of the material. Based on this information, it was shown that a relatively simple preparation process yields highly efficient and very durable anticorrosive films. They unite three essential perquisites for an appropriate coating system: a high corrosion resistance, long-term stability, and environmental compatibility. Considering also the simplicity of the sol-gel process and the low material consumption, which scales with the film thickness, these about 5-μm thick hybrid films constitute from the economical and environmental point of view a very interesting alternative for conventional protective coating systems.

Hybrid	Synthesis	Substrate	Deposition/ thickness (μm)	EIS: ∣Z∣ (GΩ cm²), lifetime (days), solution	Polarization I_{corr} (A cm^{-2}) E_{corr} (V), reference electrode	Reference
Epoxy-GPTMS-TEOS	Sol-gel	Carbon steel	Dip-coating/6.7	~1 42 3.5% NaCl	_ –0.08 Ag/AgCl	[This work]
Epoxy-APTES-ZnO	Sol-gel	Mg alloy AZ31	Dip-coating/~12	~1 35 0.05 M NaCl	_	[33]
Epoxy-APTES	_	Carbon steel	Spray/125	~100 21 0.1 M Na_2SO_4	_	[27]
Epoxy-SiO_2	Sol-gel	Mg alloy	Dip-coating/-	~100 7 3.5% NaCl	_	[34]
Epoxy-APTES	Solution intercalation method	Mild steel	Brush method/70–80	~10 30 3% NaCl	_	[35]
Epoxy-APTES-tetrathiol	Sol-gel	Al alloy AA2024-T3	Single blade/150	~1 350 0.5 M NaCl	_	[26]
Epoxy-polysiloxane	Commercial	Cold rolled low carbon steel	Air-less spray/70	~100 467 3% NaCl	_ –0.65 SCE	[7]
GPTMS-MTEOS-TEOS	Sol-gel	Al alloy AA2024-T3	Dip-coating/25	~0.1 38 5% NaCl	_	[36]
Epoxy-APTES	_	Mg alloy AZ31	Dip-coating/14	~10 31 0.5 M NaCl	_	[37]
Epoxy-GPTMS-MTEOS/-	Sol-gel	Al alloy AA2024	Dip-coating/~8	~1 51 0.05 M NaCl	10^{-10} -0.3 Ag/AgCl	[38]

GPTMS: (3-glycidoxypropyl)trimethoxysilane; TEOS: tetraethoxysilane; APTES: aminopropyl-triethoxysilane; tetrathiol: pentaerythritol tetrakis(3-mercaptopropionate); MTEOS: methyl-triethoxysilane; SCE: standard calomel electrode.

Table 7. Principal preparation parameters and results reported for epoxy-silica coatings applied for corrosion protection of metallic surfaces, including corrosion resistance ∣Z∣, current density, I_{corr}, and corrosion potential, E_{corr}.

Hybrid/additive	Synthesis	Substrate	Deposition/ thickness (μm)	EIS: |Z| (GΩ cm²), lifetime (days), solution	Polarization I_{corr} (A cm⁻²) E_{corr} (V), reference electrode	Reference
PMMA-MPTS-TEOS/-	Radical polymerization and sol-gel	Al alloy AA2024	Dip-coating/~3	~50 more than 560 3.5% NaCl	_ −0.68 Ag/AgCl	[This work]
PMMA-MPTS-TEOS/-	Sol-gel	316L stainless steel	Dip-coating/~2	~0.01 36 3.5% NaCl	10^{-9} 0.1 Ag/AgCl	[14]
GMA-EHA-GPTMS-TEOS/-	Sol-gel	Al alloys AA1050	Spin-coating/~1	~1 21 0.1 M NaCl	_	[39]
PMMA-MPTS-TEOS/-	Sol-gel	A1010 carbon steel	Dip-coating/1.5–3	~1 18 3.5% NaCl	10^{-10} −0.3 Ag/AgCl	[22]
PMMA-MPTS-TEOS/ Ce(IV)	Radical polymerization and sol-gel	A1010 carbon steel	Dip-coating/~2	~10 304/404 3.5% NaCl	10^{-11} + 0.3 Ag/AgCl	[11]
PMMA-MPTS-TEOS/ lignin	Radical polymerization and sol-gel	A1020 carbon steel	Dip-coating/~2	~0.5 50 3.5% NaCl	_	[20]
PMMA-MPTS-TEOS/-	Radical polymerization and sol-gel	A1010 carbon steel	Dip-coating/1.5–2	~5 196 3.5% NaCl	_	[10]
PMMA-MPTSTEOS/ Ce	Sol-gel	Mild steel	Dip-coating/~26	~10 362 3.5% NaCl	10^{-12} + 0.35 SCE	[31]
Acrylic resin-SiO_2/-	Solution intercalation method	Mild steel	Brush method/75	~10 90 3.5% NaCl	_	[8]
Acrylic resin-silanol-ZnO/-	Solution intercalation method	Mild steel	Brush method/75	~10 30 3.5% NaCl	_	[40]
PMMA-MPTS-TEOS/CNTs, GO	Radical polymerization and sol-gel	A1020 carbon steel	Dip-coating/3–6	~3 211 3.5% NaCl	_ +0.58 Ag/AgCl	[18]

PMMA: poly(methyl methacrylate); MPTS: 3-(trimethoxysilyl)propyl methacrylate; TEOS: tetraethoxysilane; GMA: glycidyl methacrylate; EHA: 2-ethylhexyl acrylate; GPTMS: (3-glycidoxypropyl) trimethoxysilane;CNTs: carbon nanotubes; GO:graphene oxide; SCE: standard calomel electrode.

Table 8. Principal preparation parameters and results reported for acrylic-silica coatings applied for corrosion protection of metallic surfaces, including corrosion resistance |Z|, current density I_{corr}, and corrosion potential E_{corr}.

Hybrid/additive	Synthesis	Substrate	Deposition/ thickness (μm)	EIS: |Z| (GΩ cm^2), lifetime (days), solution	Polarization, I_{corr}(A cm^{-2}), E_{corr} (V), reference electrode	Reference
Polyurethane APTES-TEOS/-	_	AA3003 H14	Drawdown bar/75	~100 263 3.5%NaCl+0.1M HCl	_	[9]
Polyurethane-ZrO_2-SiO_2/-	Sol-gel	Carbon steel	Spray/40–55	~100 226 3.5% NaCl	_	[32]
Epoxy-polyaniline-ZnO/-	Chemical oxidative method	Carbon steel	Dip-coating/118	~1 90 3.5% NaCl	_ –0.05 (SCE)	[41]
Polyetherimide-HA/-	_	Mg alloy AZ31	Dip-coating/~4	~1 101 Hank's solution	_	[42]
Epoxy-LDH/-	_	Al alloy AA2024-T3	Spray/55	~1 18 0.05M NaCl	_	[43]
Epoxy-HS/8-hydroyquinoline	_	Al alloy AA2024-T3	Dip-coating/~25	~1 90 0.5M NaCl	_	[44]
Epoxy-$CaCO_3$	_	Al alloy AA2024-T3	Dip-coating/~30	~1 41 0.5M NaCl	_	[45]
Epoxy-ester-siloxane-urea	_	Al alloy AA2024-T3	Drop-coating/130–140	~0.1 70 3.5% NaCl	10^{-10} –0.4 SCE	[46]

APTES: aminopropyltriethoxysilane; TEOS: tetraethoxysilane; HA: hydroxyapatite; LDH: Layered double hydroxide; HS: halloysites; $CaCO_3$: calcium carbonate; SCE: standard calomel electrode.

Table 9. Principal preparation parameters and results reported for a varied of hybrid coatings applied for corrosion protection of metallic surfaces, including corrosion resistance |Z|, current density I_{corr}, and corrosion potential E_{corr}.

4. Conclusions

Structural, thermal, mechanical, and electrochemical characterization of novel epoxy-silica and PMMA-silica hybrid coatings have shown that their properties are extremely dependent of the hybrid precursors proportion, time and temperature of synthesis, and addition of fillers. After a careful adjustment of the preparation conditions, these homogeneous and transparent hybrid coatings present a defect-free very smooth surface, low porosity, a highly cross-linked silica network, excellent adhesion to the metallic substrate, elevated thermal stability, and especially an excellent anticorrosive performance. Epoxy-silica and PMMA-silica films with thicknesses of less than 10 μm exhibit a dense and highly reticulated nanostructure, resulting in enhanced thermal stability combined with high corrosion resistance and long durability in saline environment. Exceptional barrier properties, especially on aluminum alloy, were found for PMMA-silica hybrids prepared at a 8MMA:1MPTS:2TEOS molar ratio, 4 h/80°C of synthesis, and BPO/MMA molar ratio of 0.01. This coating highlights a corrosion resistance of about 50 GΩ cm^2 and a lifetime of more than 18 months in saline solution. Nanofillers have been successfully added to the PMMA-silica matrix to improve the anticorrosive performance and to reinforce the hybrid structure. Carbon nanotubes and graphene oxide incorporated into the PMMA-silica matrix resulted in a multifunctional material, which combines an excellent anticorrosive performance with improved adhesion, anti-scratch and heat-resistant properties, thus extending the application range of these coatings to abrasive environments. Furthermore, it has been shown that added Ce(IV) ions act as oxidation agents during the formation of the hybrid matrix, leading to densification process that improves the barrier property of the coatings. In addition, the active corrosion inhibition provided by formation of insoluble cerium species in regions affected by corrosion, known as self-healing ability, resulted in a prolonged the lifetime of the coatings. The great progress achieved in the last couple of years in the development of organic-inorganic hybrids makes these materials very promising candidates for new-generation high-performance protective coatings.

Acknowledgements

We would like to thank the National Laboratory of Synchrotron Light Source (LNLS, Brazil) for the use of SAXS facilities. This work was supported by the Brazilian funding agencies CNPq, CAPES, and FAPESP.

Author details

Samarah V. Harb, Andressa Trentin, Ruben F. O. Torrico, Sandra H. Pulcinelli, Celso V. Santilli and Peter Hammer*

*Address all correspondence to: peter@iq.unesp.br

Instituto de Química, UNESP-Universidade Estadual Paulista, Araraquara, SP, Brazil

References

[1] Brasunas A. de S., Delinder L. S. V. Corrosion basics: An introduction. 1984. Houston: TX: National Association of Corrosion Engineers. 353 p.

[2] Wint N., Vooys A. C. A., de McMurray H. N. The corrosion of chromium based coatings for packaging steel. Electrochimica Acta. 2016;**203**:326-336. DOI: 10.1016/j.electacta.2016.01.100.

[3] Schem M., Schmidt T., Gerwann J., Wittmar M., Veith M., Thompson G. E., Molchan I. S., Hashimoto T., Skeldon P., Phani A. R., Santucci S., Zheludkevich M. L. CeO_2-filled sol-gel coatings for corrosion protection of AA2024-T3 aluminium alloy. Corrosion Science. 2009;**51**(10):2304-2315. DOI: 10.1016/j.corsci.2009.06.007.

[4] Norouzi M., Garekani A. A. Corrosion protection by zirconia-based thin films deposited by a sol-gel spin coating method. Ceramics International. 2014;**40**(2):2857-2861. DOI: 10.1016/j.ceramint.2013.10.027.

[5] Quinson J. F., Chino C., Becdelievre A. M. De, Guizard C., Brunel M. Deformation capability and protective role of zirconia coatings on stainless steel. Journal of Materials Science. 1996;**31**(19):5179-5184. DOI: 10.1007/BF00355922.

[6] Sanchez C., Belleville P., Popall M., Nicole L. Applications of advanced hybrid organic-inorganic nanomaterials: From laboratory to market. Chemical Society Reviews. 2011;**40**(2):696-753. DOI: 10.1039/C0CS00136H.

[7] Echeverría M., Abreu C. M., Lau K., Echeverría C. A. Viability of epoxy–siloxane hybrid coatings for preventing steel corrosion. Progress in Organic Coatings. 2016;**92**:29-43. DOI: 10.1016/j.porgcoat.2015.12.005.

[8] Ammar S. H., Ramesh K., Vengadaesvaran B., Ramesh S., Arof A. K. A novel coating material that uses nano-sized SiO_2 particles to intensify hydrophobicity and corrosion protection properties. Electrochimica Acta. 2016;**220**:417-426. DOI: 10.1016/j.electacta.2016.10.099.

[9] Visuet E. M., Gao T., Soucek M., Castaneda H. The effect of TiO_2 as a pigment in a polyurethane/polysiloxane hybrid coating/aluminum interface based on damage evolution. Progress in Organic Coatings. 2015;**83**:36-46. DOI: 10.1016/j.porgcoat.2015.02.001.

[10] dos Santos F. C., Harb S. V., Menu M., Turq V., Pulcinelli S. H., Santilli C. V., Hammer P. On the structure of high performance anticorrosive PMMA-siloxane-silica hybrid coatings. RSC Advances. 2015;**5**(129):106754-106763. DOI: 10.1039/c5ra20885h.

[11] Harb S. V., dos Santos F. C., Caetano B. L., Pulcinelli S. H., Santilli C. V., Hammer P. Structural properties of cerium doped siloxane-PMMA hybrid coatings with high anticorrosive performance. RSC Advances. 2015;**5**:15414-15424. DOI: 10.1039/c4ra15974h.

[12] Blaiszik B. J., Kramer S. L. B., Olugebefola S. C., Moore J. S., Sottos N. R., White, S. R. Self-healing polymers and composites. Annual Review of Materials Research. 2010;**40**:179-211. DOI: 10.1146/annurev-matsci-070909-104532.

[13] Fischer H. R., García S. J. Active protective coatings: Sense and heal concepts for organic coatings. In: Hughes A. E., Mol J. M. C., Zheludkevich M. L., Buchheit R. G., editors. Active protective coatings: New-generation coatings for metals. Dordrecht, The Netherlands: Springer Science+Business Media B.V; 2016. 139-156. DOI: 10.1007/978-94-017-7540-3_7.

[14] Hammer P., Schiavetto M. G., dos Santos F. C., Benedetti A. V., Pulcinelli S. H., Santilli C. V. Improvement of the corrosion resistance of polysiloxane hybrid coatings by cerium doping. Journal of Non-Crystalline Solids. 2010;**356**(44-49):2606-2612. DOI: 10.1016/j.jnoncrysol.2010.05.013.

[15] Pepe A., Aparicio M., Ceré S., Durán A. Preparation and characterization of cerium doped silica sol-gel coatings on glass and aluminum substrates. Journal of Non-Crystalline Solids. 2004;**348**:162-171. DOI: 10.1016/j.jnoncrysol.2004.08.141.

[16] Suegama P. H., Sarmento V. H. V., Montemor M. F., Benedetti A. V., De Melo H. G., Aoki I. V., Santilli C. V. Effect of cerium (IV) ions on the anticorrosion properties of siloxane-poly(methyl methacrylate) based film applied on tin coated steel. Electrochimica Acta. 2010;**55**:5100-5109. DOI: 10.1016/j.electacta.2010.04.002.

[17] Montemor M. F., Pinto R., Ferreira M. G. S. Chemical composition and corrosion protection of silane films modified with CeO2 nanoparticles. Electrochimica Acta. 2009;**54**(22):5179-5189. DOI: 10.1016/j.electacta.2009.01.053.

[18] Harb S. V., Pulcinelli S. H., Santilli C. V., Knowles K., Hammer P. A comparative study on graphene oxide and carbon nanotube reinforcement of PMMA-siloxane-silica anticorrosive coatings. ACS Applied Materials & Interfaces. 2016;**8**:16339-16350. DOI: 10.1021/acsami.6b04780.

[19] Harb S. V., dos Santos F. C., Pulcinelli S. H., Santilli C. V., Knowles K., Hammer P. Protective coatings based on PMMA-silica nanocomposites reinforced with carbon nanotubes. In: Berber M. R., Hafez I. H., editors. Carbon nanotubes–current progress of their polymer composites. 1st ed. Rijeka, Croatia: Intech; 2016. 195-225. DOI: 10.5772/62808.

[20] Harb S. V., Cerrutti B. M., Pulcinelli S. H., Santilli C. V., Hammer P. Siloxane-PMMA hybrid anti-corrosion coatings reinforced by lignin. Surface and Coatings Technology. 2015;**275**:9-16. DOI: 10.1016/j.surfcoat.2015.05.002.

[21] Callister W. D., Rethwisch D. G.. Appendix B/properties of selected engineering materials. In: Callister W. D., Rethwisch D. G., editors. Fundamentals of materials science and engineering: An integrated approach. 2012. 4th ed. Hoboken: Wiley. pp. 835-836.

[22] Hammer P., dos Santos F. C., Cerrutti B. M., Pulcinelli S. H., Santilli C. V. Highly corrosion resistant siloxanepolymethyl methacrylate hybrid coatings. Journal of Sol-Gel Science and Technology. 2012;**63**(2):266-274. DOI: 10.1007/s10971-011-2672-8.

[23] Cole K. C., Hechler J. J., Noel D. New approach to modeling the cure kinetics of epoxy amine thermosetting resins. 2. Application to a typical system based on bis[4-(diglycidylamino)phenyl]methane and bis(4-aminophenyl) sulfone. Macromolecules. 1991;**24**(11):3098-3110. DOI: 10.1021/ma00011a012.

[24] Dahmouche K., Carlos L. D., Zea Bermudez V., SA Ferreira R. A., Santilli, C. V., Craievich A. F. Structural modelling of Eu3z-based siloxane–poly(oxyethylene) nanohybrids. Journal of Materials Chemistry. 2011;**11**(12):3249-3257. DOI: 10.1039/B104822H.

[25] Zaioncz S., Dahmouche K., Paranhos C. M., San Gil R. A. S., Soares B. G. Relationships between nanostructure and dynamic-mechanical properties of epoxy network containing PMMA-modified silsesquioxane. eXPRESS Polymer Letters. 2009;**3**(6):340-351. DOI: 10.3144/expresspolymlett.2009.43.

[26] Zadeh M. A., Van Der Zwaag S., Garcia S. J. Adhesion and long-term barrier restoration of intrinsic self-healing hybrid sol-gel coatings. ACS Applied Materials & Interfaces. 2016;**8**(6):4126-4136. DOI: 10.1021/acsami.5b11867.

[27] Díaz I., Chico B., De La Fuente D., Simancas J., Vega J. M., Morcillo M. Corrosion resistance of new epoxy–siloxane hybrid coatings. A laboratory study. Progress in Organic Coatings. 2010;**69**:278-286. DOI: 10.1016/j.porgcoat.2010.06.007.

[28] Mallakpour S., Zadehnazari A. Preparation of dopamine-functionalized multi-wall carbon nanotube/poly(amide-imide) composites and their thermal and mechanical properties. New Carbon Materials. 2016;**31**(1):18-30. DOI: 10.1016/S1872-5805(16)60001-X.

[29] Khun N. W., Troconis B. C. R., Frankel G. S. Effects of carbon nanotube content on adhesion strength and wear and corrosion resistance of epoxy composite coatings on AA2024-T3. Progress in Organic Coatings. 2014;**77**(1):72-80. DOI: 10.1016/j.porgcoat.2013.08.003.

[30] Hammer P., dos Santos F. C., Cerrutti B. M., Pulcinelli S. H., Santilli C. V. Carbon nanotube-reinforced siloxane-PMMA hybrid coatings with high corrosion resistance. Progress in Organic Coatings. 2013;**76**(4):601-608. DOI: 10.1016/j.porgcoat.2012.11.015.

[31] Mosa J., Navarro N. C. R., Aparicio M. Active corrosion inhibition of mild steel by environmentally-friendly Ce-doped organic-inorganic sol-gel coatings. RSC Advances. 2016;**6**:39577-39586. DOI: 10.1039/c5ra26094a.

[32] López D. D. A., Crespo M. A. D., Huerta A. M. T., Vela A. F., Adame J. A., Rosales H. D. Analysis of degradation process during the incorporation of ZrO2:SiO2 ceramic nanostructures into polyurethane coatings for the corrosion protection of carbon steel. Journal of Materials Science. 2013;**48**:1067-1084. DOI: 10.1007/s10853-012-6839-7.

[33] Brusciotti F., Snihirova D. V., Xue H., Montemor M. F., Lamaka S. V., Ferreira M. G. S. Hybrid epoxy–silane coatings for improved corrosion protection of Mg alloy. Corrosion Science. 2013;**67**:82-90. DOI: 10.1016/j.corsci.2012.10.013.

[34] Qiao Y., Li W., Wang G., Zhang X., Cao N. Application of ordered mesoporous silica nanocontainers in an anticorrosive epoxy coating on a magnesium alloy surface. RSC Advances. 2015;**5**:47778-47787. DOI: 10.1039/c5ra05266a.

[35] Ammar S. H., Ramesh K., Vengadaesvaran B., Ramesh S., Arof A. K. Amelioration of anticorrosion and hydrophobic properties of epoxy/PDMS composite coatings containing nano ZnO particles. Progress in Organic Coatings. 2016;**92**:54-65. DOI: 10.1016/j.porgcoat.2015.12.007.

[36] Yuan X., Yue Z. F., Chen X., Wen S. F., Li L., Feng T. The protective and adhesion properties of silicone-epoxy hybrid coatings on 2024 Al-alloy with a silane film as pretreatment. Corrosion Science. 2016;**104**:84-97. DOI: 10.1016/j.corsci.2015.11.035.

[37] Lamaka S. V., Xue H. B., Meis N. N. A. H., Esteves A. C. C., Ferreira M. G. S. Fault-tolerant hybrid epoxy-silane coating for corrosion protection of magnesium alloy AZ31. Progress in Organic Coatings. 2015;**80**:98-105. DOI: 10.1016/j.porgcoat.2014.11.024.

[38] Kozhukharov S., Kozhukharov V., Schem M., Aslan M., Wittmar M., Wittmar A., Veith M. Protective ability of hybrid nano-composite coatings with cerium sulphate as inhibitor against corrosion of AA2024 aluminium alloy. Progress in Organic Coatings. 2012;**73**(1):95-103. DOI: 10.1016/j.porgcoat.2011.09.005.

[39] Khelifa F., Druart M. E., Habibi Y., Bénard F., Leclere P., Olivier M., Dubois P. Sol-gel incorporation of silica nanofillers for tuning the anti-corrosion protection of acrylate-based coatings. Progress in Organic Coatings. 2013;**76**:900-911. DOI: 10.1016/j.porgcoat.2013.02.005.

[40] Ammar S. H., Ramesh K., Vengadaesvaran B., Ramesh S., Arof A. K. Formulation and characterization of hybrid polymeric/ZnO nanocomposite coatings with remarkable anti-corrosion and hydrophobic characteristics. Journal of Coatings Technology and Research. 2016;**13**(5):921-930. DOI: 10.1007/s11998-016-9799-z.

[41] Mostafaei A., Nasirpouri F. Epoxy/polyaniline–ZnO nanorods hybrid nanocomposite coatings: Synthesis, characterization and corrosion protection performance of conducting paints. Progress in Organic Coatings. 2014;**77**:146-159. DOI: 10.1016/j.porgcoat.2013.08.015.

[42] Zomorodian A., Garcia M. P., Moura e Silva T., Fernandes J. C. S., Fernandes M. H., Montemor M. F. Corrosion resistance of a composite polymeric coating applied on biodegradable AZ31 magnesium alloy. Acta Biomaterialia. 2013;**9**:8660-8670. DOI: 10.1016/j.actbio.2013.02.036.

[43] Zheludkevich M. L., Poznyak S. K.,Rodrigues, L. M., Raps D., Hack T., Dick L. F., Nunes T., Ferreira M. G. S. Active protection coatings with layered double hydroxide nanocontainers of corrosion inhibitor. Corrosion Science. 2010;**52**:602-611. DOI: 10.1016/j.corsci.2009.10.020.

[44] Snihirova D., Liphardt L., Grundmeier G., Montemor F. Electrochemical study of the corrosion inhibition ability of "smart" coatings applied on AA2024. Journal of Solid State Electrochemistry. 2013;**17**:2183-2192. DOI: 10.1007/s10008-013-2078-3.

[45] Snihirova D., Lamaka S. V., Montemor M. F. "SMART" protective ability of water based epoxy coatings loaded with CaCO3 microbeads impregnated with corrosion inhibitors applied on AA2024 substrates. Electrochimica Acta. 2012;**83**:439-447. DOI: 10.1016/j.electacta.2012.07.102.

[46] Okafor P. A., Beemat J. S., Iroh J. O. Thermomechanical and corrosion inhibition properties of graphene/epoxy ester–siloxane–urea hybrid polymer nanocomposites. Progress in Organic Coatings. 2015;**88**:237-244. DOI: 10.1016/j.porgcoat.2015.07.00.

Chapter 3

Co-deposited Ni-Cr-B Nanocomposite Coatings for Protection Against Corrosion-Erosion

Jorge Morales Hernández,
María de Lourdes Montoya García,
Héctor Javier Dorantes Rosales and
Joel Moreno Palmerin

Additional information is available at the end of the chapter

http://dx.doi.org/10.5772/67639

Abstract

Electrodeposition is a low-cost and low-temperature method for producing metal matrix composite coatings. The electrodeposition of Ni matrix/Ni-Cr-B particles is considered as the co-deposition of Ni-Cr-B particles in a Ni matrix, resulting in nanocomposite coatings that can offer good wear and corrosion resistance between other applications. For comparison, the electrodeposition of Ni films and their wear and corrosion evaluation were also carried out under the same conditions. Some coatings usually contain oxide or carbide particles in micrometer size and are electrodeposited in a nickel matrix; however, the use of the mechanical alloying process offers the possibility to reduce the particle size in the order of nanometers obtaining solid solutions, amorphous phases, or intermetallic compounds during the development of new alloys to be co-deposited, improving the engineering materials properties. This kind of nanocomposite can be used in industrial components with an irregular geometry exposed in aggressive environments such as the energy generation and oil industry.

Keywords: co-deposition, nanocomposite coating, erosion-corrosion resistance

1. Introduction

Surface is responsible for the serviceability of the components and the surface engineering is a multidisciplinary activity dedicated to modify the surface according to their functional necessity and where the substrate do not have the capacity. Surface characteristics for engineering applications and in our daily life are:

- Decorative and esthetic
- Thermal barrier
- Better mechanical properties
- Electronic or electrical properties
- Corrosion resistance
- Wear resistance between others

As indicated by other authors, these properties can be enhanced by microstructural changes and phase transformations (metallurgically), by deformations (mechanically), by chemical reactions, and diffusion that modify the surface chemistry or by adding coatings by physical and chemical methods [1]. Industrial plant efficiency can be affected by corrosion and wear damage in such a way that the economic losses associated with this failure mechanism can be quantified as a percentage of a country's gross domestic product annually. In some cases, the corrosion like a failure mechanism may initiate or precede to the wear failure or vice versa, depending of the medium nature; however, the combined effect of corrosion and wear can reduce the component life drastically.

The use of more alloyed steels to improve the surface hardness and corrosion resistance means to use expensive steels with respect to the carbon or low alloy steels, so that several efforts have been made for the protection against corrosion-erosion but with a low efficiency such as the use of inhibitors and the applications of cathodic or anodic currents.

Use of organic, inorganic or metallic coatings have been a good alternative for corrosion and wear protection where the proper selection depends on the accessibility, component size, environment conditions (composition, concentration, pressure, and temperature), and cost; nevertheless, the wide variety of coatings and process development for a lot of corrosive and wear environments are not enough yet, considering which exist different corrosion forms depending on the mechanism of attack that combined with the types of wear, reducing the possibility of getting the best protection.

Tribocorrosion is the concept that explains the surface degradation mechanism when mechanical wear and chemical/electrochemical processes interact with each other [2]. Many industries show damages associated with the tribocorrosion mechanism such as aeronautic, geothermic energy generation, paper industry, and making steel to name a few. To solve the combined failure mechanisms of corrosion and erosion, different alloy systems deposited with the same process have been studied in monolayers as well as multilayers; in this context, the techniques combination, which can offer a variety of coatings with the enough thickness of corrosion protection, with the homogeneous dispersion of fine particles, high hardness, and lubricant properties, has motivated the development of composite coatings where the electrodeposition techniques can offer metal matrix composite coatings, though the co-deposition of nanoparticles with superior properties at low cost.

2. Development of nanocomposite

2.1. Electrochemical nanocomposite coatings (ENC)

Electrodeposition or electroplating process is defined as the deposition of a coating by electrolysis, depositing a substance on an electrode immersed in an electrolyte by the application of electric current through the electrolyte [1]. Modification of the electroplating process includes the occlusions of metallic or nonmetallic particles dispersed in the bath to obtain a composite coating.

Oxides, carbides, silicides, refractory, metallic, and organic powders such as SiC, SiO_2, Al_2O_3, and TiO_2 have been electrodeposited with pure metals to increase strength, harness, wear, and corrosion resistance at the surface with a particle size in the order of microns and current density of 20–200 mA/cm^2 with agitation. The incorporation of a homogeneously dispersed second phase in a metal matrix offers new engineering properties. During the co-deposition, small metal particles added to the plating bath are deposited or embedded simultaneously into a metallic matrix during the electrolytic process [3]. The properties of the composite coating depend on the nature, distribution, and size of the particles in combination with the metallic matrix. The development of a lot of alloying systems by powder metallurgy offers to the composite coating technology, the alternative to increase the opportunities in the solutions of corrosion and wear failures.

Commercial electrolytic baths such as nickel, chromium, copper, and zinc are the metallic matrix, and the dispersed phase is an insoluble solid with nanometric particle size and known morphology, which remains on the cathode surface by agitation and overgrown with the metal electrodeposition [4]. Since the 1970s, the electrodeposited composite coatings have been improved and have been published in some theoretical models that describe the adsorption and electrophoretic migration of particles, until the incorporation of a corrective factor to account for the effects of adsorption and hydrodynamic conditions [5].

Figure 1 reported by Low et al. shows the general process of the co-deposition of particles into decoating, which include: (i) formation of ionic clouds around the particles, (ii) convective movement toward the cathode (convection layer), (iii) concentration boundary layer (diffusion layer), (iv) electrical double layer followed by (v) adsorption and encapsulated of particles. Some theories consider only the transport of particles due to electrophoresis, mechanical entrapment, adsorption, and convective diffusion [5].

Additional to the typical process variables in electroplating such as bath constituents, temperature, pH, current density, agitation, and surfactants is necessary to consider the variables that toward the name of electrochemical co-deposition to obtain composite coatings; these parameters are concentration, size, type, and shape of the particles between others.

Microstructure of the particles can modify the kinetic in such a way γ-Al_2O_3 deposition with copper requests less particle concentration than α-Al_2O_3 [6]. The influence of the particle size and shape are associated with the surface relation, affecting the adsorption of the particles on the cathode, the adsorption of ions on the particle surface, and the suspension stability

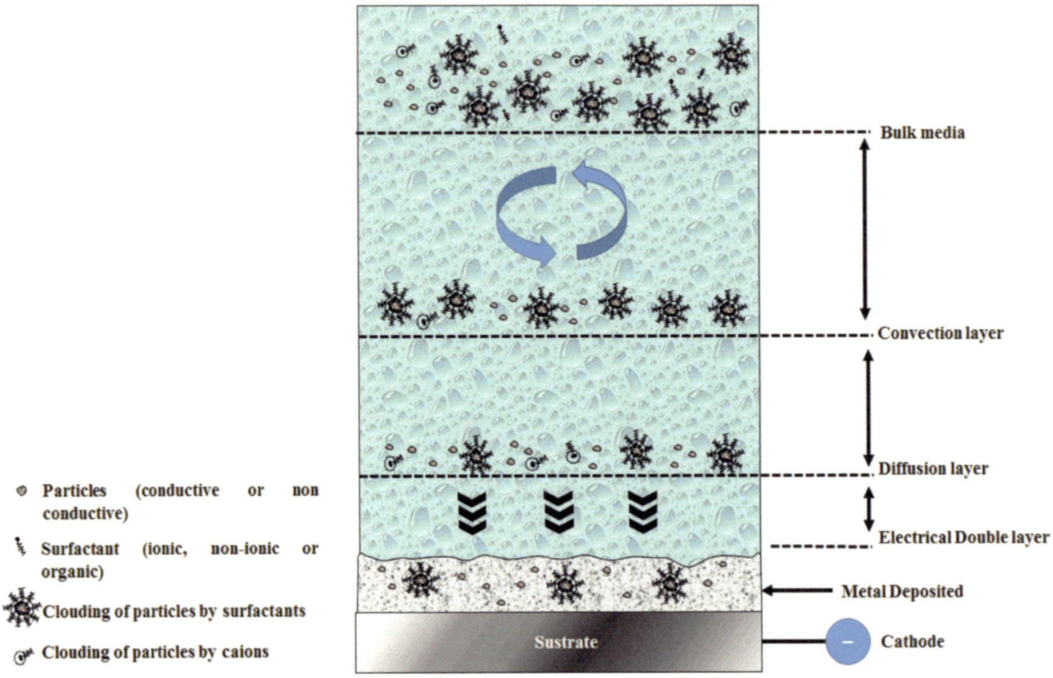

Figure 1. Process involved in the co-deposition of particles into metallic coating [5].

during the co-deposition process. Temperature have a great influence in each co-deposition systems, for example, Ni-Al_2O_3 is not affected by the temperature on the percentage of embedded particles; on the other hand, the density of particles in the system Ni-Cr increased with the temperature up to 50°C [7].

2.2. Nickel matrix nanocomposite

Nickel matrices electrochemical nanocomposite coatings are characterized by the high hardness, corrosion–erosion resistance, and good appearance and which depends on the dispersed particles whose affinity with nickel could be correlated with their position in the electromotive force series (emf).

Diamond particles with a spherical shape and a concentration of 30 g/L in the bath were electrodeposited in a nickel matrix. Differences in the diamond to nondiamond ratio of carbon forms modified the adsorption characteristics and the quantity of diamond particles in coating (0.2–1 wt%). Largest density of particles into the nickel coating was achieved with the minimum diamond to nondiamond ratio, improving wear resistance and microhardness from 250 to 440 kg/m^2 with respect to pure nickel coating [4]. Hard materials like WC and SiC in a nickel matrix are used in engine block of aluminum alloys with good results in abrasion protection.

The expansion of the electrochemistry of composite materials to new areas includes application in electrocatalysis where hydrophobic Ni + PTFE composite electrodes for electrochemical reactions of water soluble organic substrates reported the evolution of oxygen and hydrogen at a lower overpotential on Ni + PTFE than on pure Ni. Combinations with $LaNiO_3$, RuO_2, metallic powders, TiO_2, and FeS reported that the most efficiency in catalysis was observed with the Ni + RuO_2 composite. The development of composites for photoelectrocatalysis applications consists of an electrochromic matrix and a semiconducting dispersed phase. In this case, the semiconductor particles are optically excited and electrons are transferred from the conduction band to the electrochromic material, which change color because of redox reaction. Secondary lithium batteries have been investigated with polypyrrole (Ppy) matrix with high concentration of MnO_2 particles [6].

Composite coatings for dry self-lubricating surfaces in automotive and industrial applications were studied by the co-deposition of graphite and MoS_2 nanoparticles in a nickel matrix, obtaining a low friction coefficient of 0.5 and with the double the wear resistance with respect to the steel substrate. CrAlY powder in cobalt or nickel matrix have been co-deposited with better results with respect to the same coating deposited with plasma sprayed because of less porosity. Aluminum particles co-deposited in a nickel matrix and heat treatment reported a nickel–aluminum intermetallic coating with excellent oxidation resistance [1].

2.3. Nanomaterials by mechanical alloying

Scientists and engineers have developed alternative processes to synthesize new materials that by the conventional methods are expensive or complicate. Rapid solidification process (RSP) and mechanical alloying (MA) are two such processing methods to produce equilibrium and metastable phases. MA was implemented to produce oxide dispersion strengthened (ODS) nickel and iron based superalloys for applications in the aerospace industry in 1966 [8].

MA is a dry powder processing technique involving cold welding, fracturing, and rewelding between clean surfaces in contact with elemental or combined powders exposed to high-energy collision in a high-energy ball mill. Development by Benjamin to produce an oxide dispersion-strengthening alloy with γ' precipitation hardening in nickel-based superalloys intended for gas turbine applications. The alloying of immiscible elements is possible by MA, increasing the solubility limit with respect to the equilibrium diagrams.

Elemental powders mixture is loading with grinding medium (ceramic or hardened steel) in a stainless-steel container sealed after be exposed at vacuum cycle and under inert atmosphere to avoid reaction with the ambient. Usually, 1–2 wt% of a process control agent (stearic acid or alcohol) is normally added to maintain an equilibrium between the welding and fracture events, especially when ductile–ductile mixture is milled. The high-energy mills are SPEX mill (10 g of powder), attritors (few pounds of powder), or planetary mills with two or more containers can be processed at a time. Since, the 1980s have been synthesized many alloy phases including equilibrium and supersaturated solid solutions, crystalline and quasicrystalline intermediate phases, and amorphous or glassy alloys [8]. Reduction of grain size of few nanometers (<100 nm) of powder mixtures by MA introduced the concept of nanocrystalline materials before the concept of nanotechnology. The research, technology development, and innovations in MA have applications in different industrial requests as is shown in **Figure 2**.

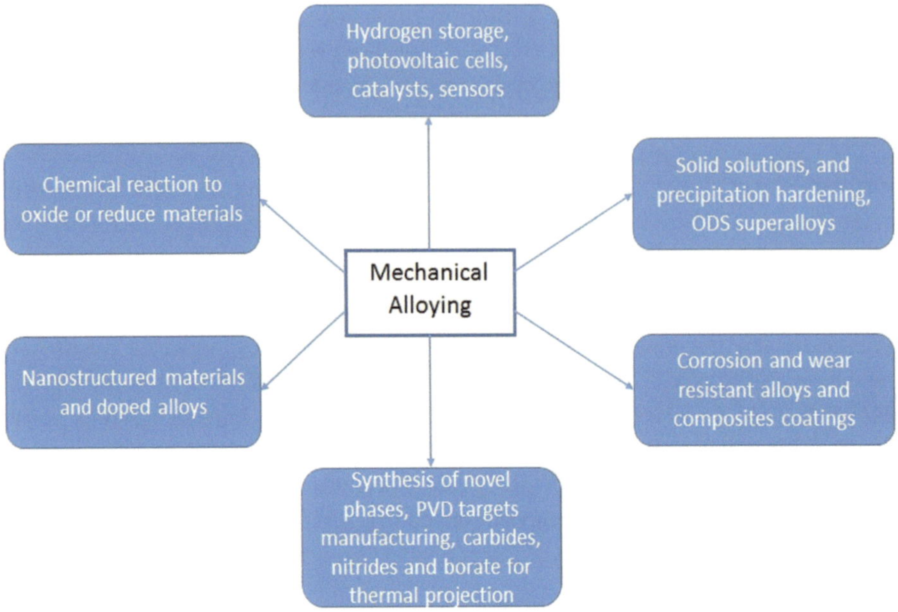

Figure 2. Potential applications of mechanically alloyed products [8].

During the particles reduction by fracturing and rewelding of the powders, the reaction area increases and the diffusion distance between particles is minimized with the introduction of many crystal defects as a function of the time during MA [9]. Thermal energy of the particles produce that all the atoms vibrate about their equilibrium position, resulting in a jump of an interstitial atom to an adjacent interstice after the atoms of the parent lattice must be forced apart into higher energy position. Activation energy for diffusion is equal to the sum of the activation energy to form a vacancy and the activation energy to move the vacancy, written as:

$$\Delta Q = \Delta Q_f + \Delta Q_m \qquad (1)$$

where ΔQ_f is the activation energy for creating vacancies and ΔQ_m the activation energy for moving vacancies with the temperature increase that is produced during the collisions between particles until it reaches the diffusion temperature.

Energy stored by the accumulate deformation during MA is associated with the creation of dislocations and grain boundaries in the deformed material, if there is sufficient activation energy by deformation process, interstitial and substitutional atoms can move in crystal lattices from one atomic site to another, promoting the diffusion process. The formation of large number of defects (dislocation and vacancies) generated by thermal energy in conventional process in MA is active by mechanical deformation during the high-energy collision between powder particles, being the activation energy for the diffusion that obey the first law of Gibbs of diffusion written as:

$$D = D_o \exp\left[\frac{-(\Delta Q_f + \Delta Q_m)}{RT}\right] \qquad (2)$$

where D is the diffusion coefficient, D_o is the diffusion coefficient where $T = 0$, R is the universal gas constant, and T is the temperature in Kelvin.

Equation (2) establishes that at the same value of D, a decrease in activation energy is equivalent to an increase in temperature, situation that is present punctually during the particle size reduction and the increase of surface energy where the localized temperature can increase considerably. It is believed that lowering activation energy is present in the MA process [9]. In the melting process, the diffusion is controlled by thermal energy; in mechanical alloying, formation of new alloys is determinate by both thermal and mechanical energies. By creating nanometer crystalline through repeated fracturing and cold welding of powder particles, diffusion take place through the grain boundaries, increasing the solubility limits inclusive in difficult alloying systems.

2.4. Ni-Cr-B nanocomposite coatings

2.4.1. Experimental

Mechanical alloying process was implemented to synthetize nanoparticle with a nominal composition of Ni-20Cr-10B (wt%) from elemental Ni, Cr, and B of 99.95, 99.80, and 99.5% purity, respectively. The powders were loaded in steel vials with hardened balls of 4.76 and 12.7 mm in diameter. The charged vial was evacuated with a vacuum pump for 20 min and then filled with argon gas in a glove box. Ethyl alcohol (1.5 c.c.) was used as the process control agent (PCA) to prevent the agglomeration of powders during milling. The high-energy equipment used was the SPEX mill 800 with a powder to ball weight ratio of 6:1 and a total sample weight of 28 g. The powders were processed until complete 40 h of milling. The phase structure of milled powders was characterized with X-ray diffraction (XRD) using a Bruker D8 advanced diffractometer with Cu Kα radiation ($\lambda = 1.54056$ nm) and operated at 40 kV and 40 mA. Powder morphology and particle size (maximum milling time) were characterized by using a JEOL 2000 scanning electron microscopy (SEM). Validation of particle size was made with a Zeta Sizer equipment model Malvern MPT-2.

Ni-Cr-B powders synthetized by MA and with average particle size of 95 nm in diameter were added in Ni-matrix with the electrochemical deposition process from a nickel sulfate bath containing 270 g/L $NiSO_4 \cdot 6H_2O$, 50 g/L $NiCl \cdot 6H_2O$, 35 g/L H_3BO_3 and 2 g/L of Ni-Cr-B as nanoparticles dispersed under constant magnetic agitation (600 rpm). The pH value of the bath was 4.3 after loading the Ni-Cr-B nanoparticles at constant temperature of 328 K (55°C). Low carbon steel plate with size of 50 mm × 20 mm × 2 mm was used as substrate after abraded it with SiC paper of grades 320, 430, 600, and 1000 and cleaned with ultrasonic in distilled water. Ni plate (99.9% purity) with similar dimensions to the steel was used as a cathode. The current density of 5.8 A/dm^2 was applied during 40 min. For comparison, nickel coating was also deposited with the same parameters and the same bath but without adding of Ni-Cr-B particles.

Surface morphology of the composite was characterized by SEM; cross-sectional was prepared to measure the coating thickness in function of deposition time by optical microscopy. Measurements of the Vickers microhardness were performed on the cross-sectional by using a microhardness tester under a load of 0.45 N for 15 s according to the ASTM E140 standard; the final reported value was the average of 10 measurements.

The friction and wear test were performed at room temperature without lubricant on a ball-on-disc type tribometer with a constant rotation speed of 200 rpm, a constant radius of 2.5 mm and load of 2 N. Hardened steel balls of 10 mm of diameter were used; the test lasted for 1 h for a total distance of 188 m and the value reported was an average of three measurements. After wear test, the worm surfaces were evaluated with optical microscopy.

Corrosion resistance of the coatings was evaluated using the anodic polarization method. The equipment used was a potentiostat/galvanostat BioLogic® with EcLab software. Open circuit potential was monitored until equilibrium was reached and then, the polarization technique was applied with an over potential of ±1 V with a scan rate of 1 mV/s. A typical three electrodes configuration with composite coating as working electrode (WE), platinum mesh as counter electrode (CE), and calomel as reference electrode (RE) were used and immersed in NaCl solution at 3.5 wt% prepared with distilled water [11]. Optical microscopy was used to evaluate the corrosion mechanism in the nickel and composite coatings and their comparison with the corrosion observed at the substrate.

2.4.2. Results and discussion

2.4.2.1. Mechanical alloying of powders

After 40 h of milling, the XRD characterization shows a solid solution (SS) of chromium and boron in nickel matrix (**Figure 3**) with a minimum particle size of 95 nm immersed in an acicular morphology. Initial powder particles of nickel change from angular to acicular powder particles (**Figure 4**) because of the introduction of two dense structures (bcc from Cr and hcp

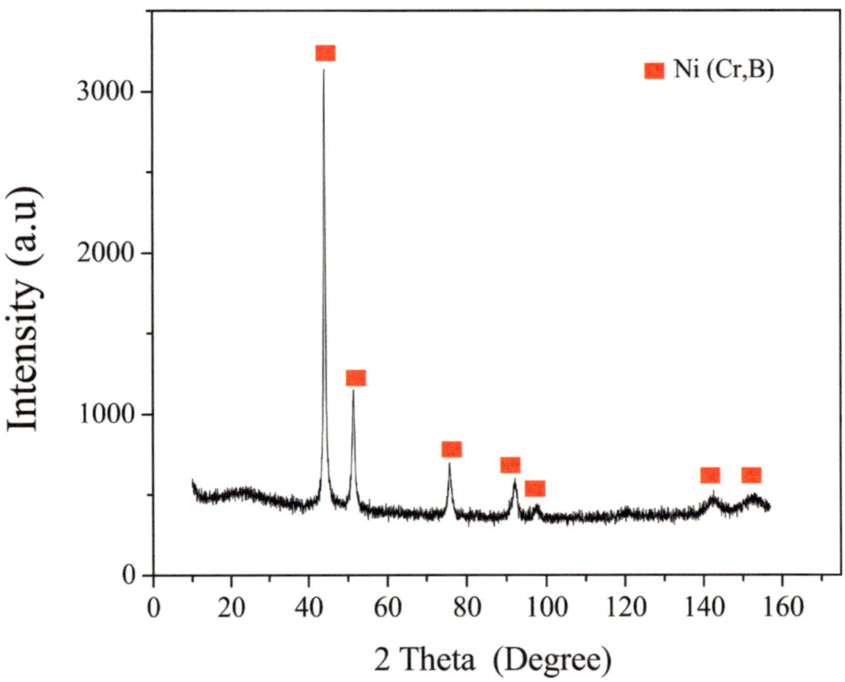

Figure 3. XRD pattern of the mixture Ni-20Cr-10B after 40 h of MA.

Figure 4. Initial morphology of Ni powders and acicular morphology in the SS after 40 h of milling.

from B) [10]. High ductility of nickel permits that two elements with similar atomic radio can be introduced in their structure. The final powder showed a high reactivity because of the high reduction in particle size and to the high stored energy by deformation.

2.4.2.2. Surface morphology in the coating

Table 1 shows the nickel electroplating thickness and the composite coating thickness at different process time, where the thickness reported was of 13.749 and 22.018 μm to the nickel and composite coatings, respectively at maximum process time. The difference in thickness with or without particles corresponding with the dispersion and over saturation of particles at the coating, resulting in a difference of 8.269 μm in thickness at 2400 s, that can mean the particles deposition without being encapsulated by the Ni matrix. During the first two steps, particles deposition was not detected in the Ni coating.

Figure 5 shows the typical morphology of the nickel electroplating that correspond with a pyramidal morphology and is reported that the surface morphology of the composite coating with 2100 s of deposition, where is observed a homogeneous dispersion of particles that in more proportion are encapsulated in the Ni matrix, identifying some coarse particles deposited over the first encapsulated particles.

When the electrodeposition time is enough, the steeps of boundary layer, electrical double layer, and the encapsulated particles are reduced by the oversaturation of particles that were

Time (s)	Identification code	Thickness (μm)	
		Ni	Ni–Ni-Cr-B
1200	–	7.033	–
1500	–	9.427	–
1800	P1	9.629	14.357
2100	P2	10.365	15.843
2400	P3	13.749	22.018

Table 1. Average thickness of Ni coating and Ni–Ni-Cr-B composite coating vs process time.

transferred in the last two steps (formation of ionic clouds and convective movement) [12]. For these reasons, some particles without being encapsulated are deposited over the first covered particles at the Ni matrix [13]. **Figure 6** shows the composite coating deposited at 2400 s where some conglomerate particles are deposited over the first layer of particles covered by nickel. This second layer shows coarse particles that could be with less adherence at

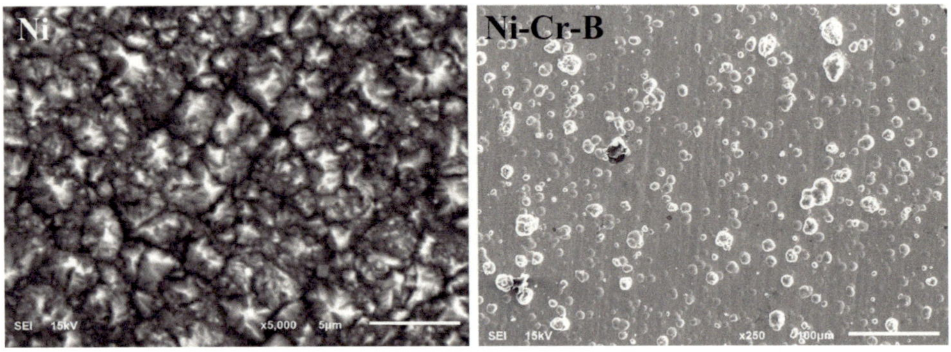

Figure 5. Nickel electroplated surface and Ni–Ni-Cr-B composite coating at 2100 s of process respectively.

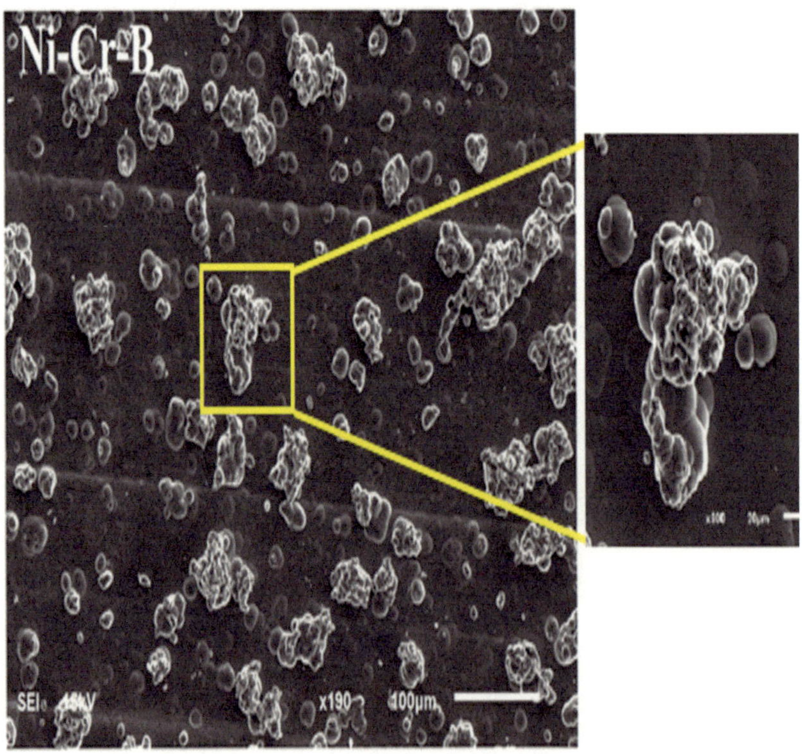

Figure 6. Composite coating deposited at 2400 s where some particles are conglomerates.

the substrate and be removed under abrasive wear conditions; however, this situation was not observed.

2.4.2.3. Hardness and wear resistance

Figure 7 shows the hardness from the samples P1, P2, and P3 in comparison with the nickel coating and substrate. Hardness composite coating is approximately 60% upper than the electroplating Ni coating and much more than the substrate. This hardness value is in average similar to obtained by heat treatment in an alloyed steel (59 HRc) for automotive and wear applications.

Friction coefficient from the substrate, Ni coating, and Ni–Ni-Cr-B composites are reported in **Figure 8**. Lubricant condition of the composite coating is observed with the minimum friction coefficient of 0.5 with respect to the Ni plating of 0.64 and carbon steel of 0.72. Differences in the co-deposition time for the composite coatings (samples P1, P2, and P3) do not represent changes in the friction coefficient independently that exist various particles density at the surfaces or in the case where exist an overdeposition of particles or conglomerates at the maximum process time, indicating the friction coefficient reported represents the composite coating nature.

Width of friction marks and appearance shows an adhesive wear condition for the Ni plating and composite coatings [14]; however, less footprint width was observed for the composite

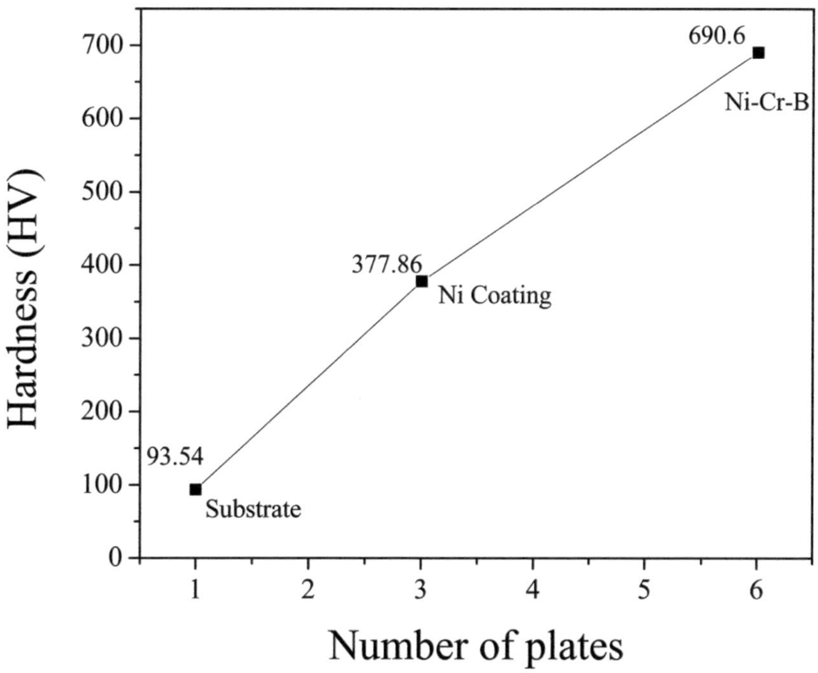

Figure 7. Vickers hardness of the composite coatings from the samples P1, P2, and P3 with respect to the Ni coating and substrate hardness.

coatings as is shown in **Figure 9**. Summarized of tribology properties is shown in **Table 2**, where the lowest friction coefficient and highest hardness reported in the composite coatings, represents a coating that can be used under abrasive operation conditions.

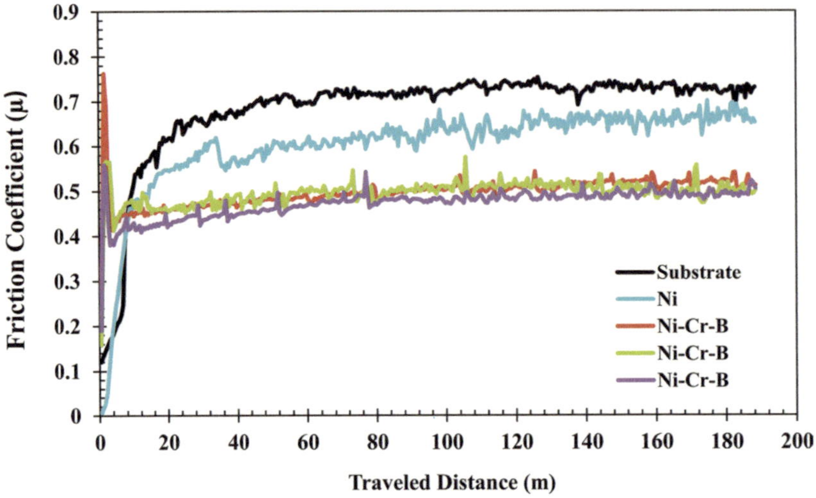

Figure 8. Friction coefficient of carbon steel, pure Ni, and Ni–Ni-Cr-B composite coatings (samples P1, P2, and P3).

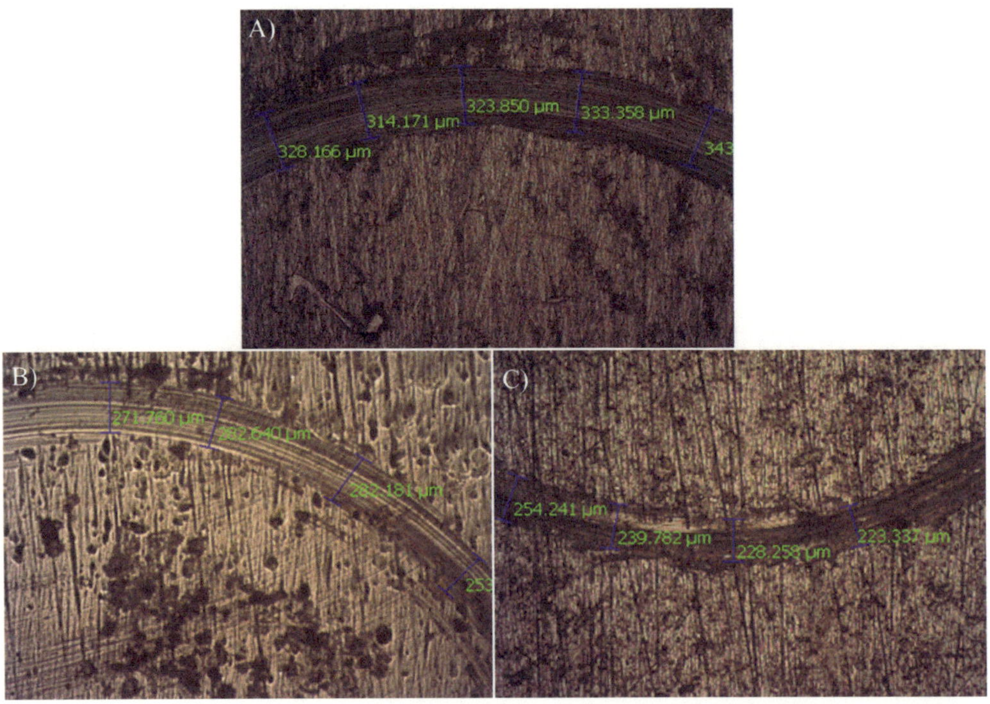

Figure 9. Footprint width for the coatings after evaluating their wear resistance for: (A) carbon steel substrate, (B) Ni plating coatings, and (C) Ni–Ni-Cr-B composite coating.

Sample	Friction coefficient	Footprint (μm)	Hardness (Hv)
Substrate	0.72	506	93.54
Ni coating	0.64	284.47	377.86
Ni–Ni-Cr-B composite	0.50	213.12	690.6

Table 2. Summarize of the physical properties from the composite and Ni coatings.

2.4.2.4. Corrosion resistance

Polarization plots show that the corrosion resistance of the Ni and Ni-Cr-B composites coatings are slightly similar, but with a higher resistive behavior for the composite coating when moving toward less current density as is reported in **Figure 10**. Minimum passive region in the anodic area was observed for Both coatings. Corrosion resistance in millimeters per year (mmpy) is reported in **Table 3**, where the composite coating reported good corrosion resistance; however,

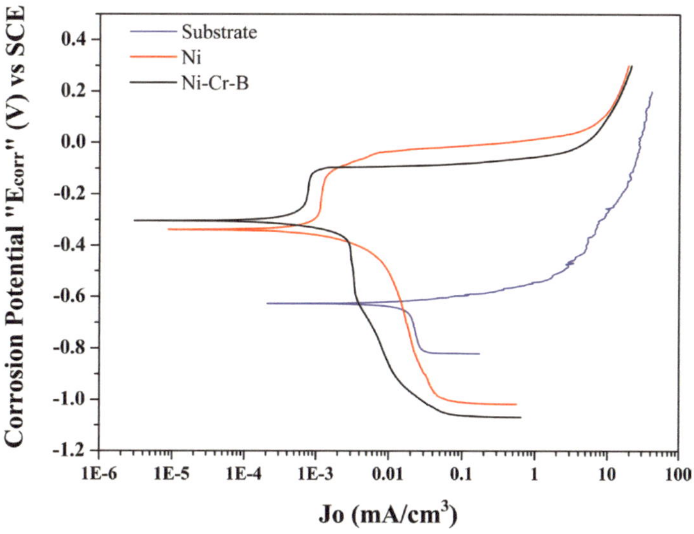

Figure 10. Polarization curves of substrate, Ni plating, and Ni–Ni-Cr-B composite coating, tested in an electrolyte of 3.5 wt% of NaCl.

Sample	V_{corr} (mmpy)
Substrate	2.120
Nickel	0.0093
Ni-Cr-B P1	0.0032
Ni-Cr-B P2	0.0037
Ni-Cr-B P3	0.0037

Table 3. Corrosion resistance in mmpy in the Ni plating and composites coatings.

due to the matrix Ni, the concentration of Ni-Cr-B particles, size, and morphology, not represent a significative difference in the corrosion resistance with respect to nickel coating, being evident the improvement in the abrasión resistance. The general corrosion mechanism observed in the substrate in NaCl was reduced in high magnitude in the nickel matrix composites.

3. Conclusions

Solid solutions of Cr and B in Ni were obtained after 40 h of MA from the elemental powders with a minimum particle size of 95 nm and with an acicular morphology. Due to the high ductility of the mixture and the high deformation energy obtained at the maximum milling time, the nanoparticles were grouped in conglomerates of the order or microns. Different coating thickness were obtained in the Ni plating and composites coatings in function of process time, obtaining a differential in thickness between the Ni pure and Ni-composite in the order of 5.47–8.26 microns and which corresponds with the increase of thickness added by the Ni-Cr-B particles, representing a more dispersion and concentration of particles proportional with the deposition time. High hardness and better wear resistance were obtained in all the composite samples independent of the process time and the particles concentration at the surface, resulting a composite coating with low friction coefficient for lubricant applications and high hardness for wear requirements. Corrosion resistance was improved in the Ni-Cr-B composites coatings lightly with respect to the Ni coating assuming that the composite coating was conformed of a Ni matrix and that the presence and concentration of Ni-Cr-B nanoparticles were not representative to improve the corrosion resistance with respect to Ni coating. We are considering the application of a heat treatment at the composite coating to increase the corrosion resistance.

Acknowledgements

The authors wish to thank our institutions (CIDETEQ, ESIQIE and UAG) for the facilities in all time and the National Council of Science and Technology (CONACYT) for the financial support through the Researchers National System (SNI).

Author details

Jorge Morales Hernández[1]*, María de Lourdes Montoya García[1], Héctor Javier Dorantes Rosales[2] and Joel Moreno Palmerin[3]

*Address all correspondence to: jmorales@cideteq.mx

1 Research Center and Technology Development in Electrochemistry, Querétaro, México

2 School of Chemical Engineering and Extractive Industries, Mexico City, México

3 Autonomous University of Guanajuato, Guanajuato, México

References

[1] J.R. Davis; Surface Engineering for Corrosion and Wear Resistance; ASM International, 2005; pp. 1-159.

[2] R.J.K. Wood; Tribo-corrosion of coatings: A review; Journal of Physics D; 40, 2007; pp. 5502-5521.

[3] J.R. Roos, J.P. Celis, J. Fransaer and C. Buelens; The development of composite plating for advanced materials; Journal of the Minerals, Metals and Materials Society; 42, 11, 1990, pp. 60-63

[4] V.N. Tseluikin; Composite electrochemical coatings: Preparation, structure, properties; Protection of Metals and Physical Chemistry of Surfaces; 45, 3, 2009, pp. 287-301.

[5] C.T.J. Low, R.G.A. Wills, F.C. Walsh; Electrodeposition of composite coatings containing nanoparticles in a metal deposit; Surface & Coatings Technology; 201, 2006, pp. 371-383.

[6] M. Musiani; Electrodeposition of composites: An expanding subject in electrochemical materials science; Electrochemica Acta; 45, 2000, pp. 3397-3402.

[7] A. Hovestad, L.J.J. Janssen; Electrochemical codeposition of inert particles in a metallic matrix; Journal of Applied Electrochemistry; 25, 1995, pp. 519-527.

[8] C. Surynarayana, E. Inavov, V.V. Boldyrev; The science and technology of mechanical alloying; Materials Science and Engineering A; 304-306, 2001, pp. 151-158.

[9] L. Lu and M.O. Lai; Formation of new materials in the solid state by mechanical alloying; Materials & Design; 16, 1, 1995, pp. 33-39.

[10] O.D. Neikov, S.S. Naboychenko, I.V. Murashova, V.G. Gopienko, I.V. Frishberg and D.V. Lotsko, Eds.; Handbook of None-ferrous Metal Powders, Technology and Applications, first edition; Elsevier, 2009.

[11] A. Mandujano Ruiz, J. Morales Hernández, F. Castañeda Saldivar, H. Herrera Hernández and J.M. Juárez García; Determination of the Kinetic Parameters for Stainless Steel 304 Exposed in Chloride Solution from Electrochemical Noise; in review to be published.

[12] S. Srikomol, Y. Boonyongmaneerat and R. Techapiesancharoenkij; Electrochemical codeposition and heat treatment of nickel-titanium alloy layers; Metallurgical and Materials Transactions B, 44B, 2013, pp. 53-61.

[13] H.-K. Lee, H.-Y. Lee, J.-M. Jeon; Codeposition of micro and nano-sized SiC particles in the nickel matrix composite coatings obtained by electroplating; Surface and Coatings Technology; 201, 8, 2007, pp. 4711-4717.

[14] Z. Yue-bo, Z. Guo-gang, Z. Hai-jun; Fabrication and wear properties of co-deposited Ni-Cr nanocomposite coatings; Transactions of Nonferrous Metals Society of China; 20, 2000, pp. 104-109.

Chapter 4

Protective Coatings for Low-Cost Bipolar Plates and Current Collectors of Proton Exchange Membrane Electrolyzers for Large Scale Energy Storage from Renewables

Philipp Lettenmeier, Aldo S. Gago and K. Andreas Friedrich

Additional information is available at the end of the chapter

http://dx.doi.org/10.5772/intechopen.68528

Abstract

Hydrogen produced by proton exchange membrane (PEM) electrolysis technology is a promising solution for energy storage, integration of renewables, and power grid stabilization for a cross-sectoral green energy chain. The most expensive components of the PEM electrolyzer stack are the bipolar plates (BPPs) and porous transport layers (PTLs), depending on the design. The high cost is due to the fact that the employed materials need to withstand corrosion at 2 V in acidic environment. Currently, only titanium is the material of choice for the anode side. We use vacuum plasma spraying (VPS) technology to apply highly stable coatings of titanium and niobium to protect stainless steel BPPs from the oxidative conditions on the anode side. The latter is able to decrease the interfacial contact resistance and improves the long-term stability of the electrolyzer. Furthermore, porous transport layers (PTL) can be realized by VPS as well. These coatings can be produced on existing titanium current collectors acting as macro porous layers (MPL). Lastly, free standing multifunctional structures with optimized tortuosity, capillary pressure and gradient porosity are used as current collectors. The coatings and porous structures developed by VPS enable the reduction of the required material and costs without performance losses.

Keywords: coatings, PEM electrolysis, PEM electrolyzers, bipolar plates, cost, current collectors, gas diffusion layer, GDL, hydrogen production

1. Introduction

On November 4, 2016, the Paris Agreement entered into force aiming to limit the global temperature rise to at least 2°C above the pre-industrial level [1, 2]. To achieve this goal, the necessary worldwide net zero carbon emission point is expected to be reached between 2045 and 2060 [3]. The energy sector represents worldwide the biggest greenhouse emitter but thanks to the resent progress in renewable energy technology such as wind, water and solar, the total energy consumption is to be substituted by green energy. However, those energy sources are strongly fluctuating and difficult to control increasing the need of large-scale energy storage. Independently of geological environment, water electrolysis is a promising technology to convert electricity to chemicals such as hydrogen and oxygen by splitting water. Hydrogen is the basis of all relevant energy carriers and enables even the connection of different sectors such as mobility or industry, the two other main greenhouse gas emission sources.

Already in 1800, William Nicholson and Anthony Carlisle established a new field in chemistry by splitting water by using direct current, that is, the electrochemistry [4]. It was Russell and co-workers who published first in 1973 the use of a solid polymer electrolyte (SPE) and anticipate the huge potential for a future energy market [5]. The two main technologies for a sustainable hydrogen production are Alkaline- and polymer exchange membrane (PEM) electrolyzers. Alkaline electrolysis is a well-established and mature technology. However, based on efficiency [6, 7], flexibility [8] and power density [6, 9, 10], the potential of economic hydrogen production by PEM electrolysis is higher, which justifies the increased interest in this technology even if it is more costly. On the other hand, open questions regarding durability and degradation remain. Moreover, the needs of rare and expensive metal, which are required to withstand the harsh acidic conditions, delay the large-scale penetration of PEM electrolyzers in the market. Indeed, potentials up to 2 V, pH values between 2 and 0 in oxygen-saturated environment require outstanding properties of the used materials.

Currently, there is no PEM electrolyzer supplier who does not use iridium as an oxygen evolution reaction (OER) catalyst, which is the rarest metal on earth. However, it is not the electro catalytic material the one that dominates the production costs of the PEM technology. In fact, the metallic parts such bipolar plates and porous transport layers are the most expensive components of a PEM electrolyzer stack. The main part of a PEM electrolyzer system is the stack consisting of several cells. Each cell consists of an anode (oxygen evolution reaction, OER) and a cathode (hydrogen evolution reaction, HER) separated by an acidic proton conductive membrane. **Figure 1** presents a scheme illustrating the working principle of PEM electrolyzers as well as the internal components. In most cell designs, the electrodes are attached directly to the proton exchange membrane. This membrane electrode assembly (MEA) is the core component of a PEM cell. Current collectors, also called porous transport layers (PTL), on both sides of the MEA are permeable to water and the product gases, allowing electric current to flow to and from the electrodes. The two half-cells are surrounded by bipolar plates (BPP), which have usually flow fields. Their function is to transport the reactant water to the membrane-electrode interface and remove the product gases.

Depending on design, the stack accounts for up to 60% of the overall system cost [11]. The PTL and BPP can be defined as interconnectors and correspond to 50–70% of the stack costs

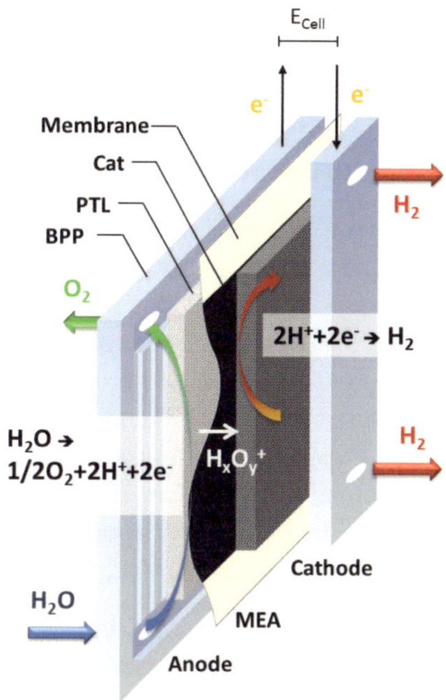

$$Anode: H_2O \rightarrow \frac{1}{2}O_2 + 2H^+ + 2e^-$$

$$Cathode: 2H^+ + 2e^- \rightarrow H_2$$

$$\text{Reaction: } H_2O \rightarrow H_2 + \frac{1}{2}O_2$$

Figure 1. Scheme of a PEM electrolyzer cell. The anode side is filled with water, which diffuses through the PTL to the iridium electrode. The liquid is subsequently spitted into O_2, $4e^-$ and $4H^+$ by theoretical potentials >1.23 V. The protons are transported to the cathode side by the PEM and combine at the cathode with the electrons from the external circuit forming hydrogen gas.

[7, 11]. These require stable metals mainly on the anode side of the electrolyzer, which is the electrode that splits water into protons, electrons and oxygen. Titanium is the state of the art material for manufacturing BPPs and PTLs. A thin layer of TiO_x passivates the metal protecting it from further degradation and corrosion. However, the material is costly and difficult to manufacture. Furthermore, the semi-conductive behavior of titanium oxides decreases the efficiency requiring often the use of a protective coating to decrease the contact resistance and prevent the oxidation of titanium.

In this chapter, we introduce the reader into the possibility of reducing the predominant asset of the investment cost for PEM electrolyzer by using protective and easily up-scalable coating technologies. Vacuum plasma spraying, a versatile applicable technology to apply various types of coatings to a wide range of surfaces, is used to produce highly stable and multifunctional coatings for cost-effective interconnectors of PEM electrolyzers.

2. Protective and multifunctional coatings for PEM electrolysis

2.1. Vacuum plasma spraying (VPS)

The Swiss engineer Max Ulrich Schoop can be considered as the inventor of the thermal spaying technology which he patented in 1909. Compared to electro/electroless plating, chemical vapor deposition (CVD) or physical vapor deposition (PVD) coating techniques, the "thermal spray" family enables the production of thick layers of several tens of microns with controlled porosity at scalable production rates. Several metallic and ceramic powders between some nanometers and several hundreds of micrometers can be used as spraying material. Interestingly, the plasma spraying technology is very suitable for the production of low-cost BPPs based on stainless steel substrate and PTLs due to the mechanical stability and relevant thickness of the produced layer [12].

The main part of this technology is the torch, where the gas flows through the annular gap of a finger like cathode and a concentrically surrounding anode. The gas consists of a mixture of Ar, N_2, H_2 and He, and it becomes ionized by electric arcs between the electrodes of several hundred amperes and heats it up to at least 10,000°C. The heat enables the complete or semi melting of the powder and accelerates it in to direction of the substrate [14]. The plasma enthalpy, the powder injection technology and particle size of the material affect this process. A completely melted particle is ideal for producing dense and protective layers. Conversely, partly melted particles can be used for manufacturing porous coatings for multiple purposes. **Figure 2a** and **b** shows a scheme of VPS spraying technology and photo of the plasma torch, respectively.

Aerospace application dominates the use of thermal spry technology, but other industries and especially the automotive sector show an increased interest. An extended overview of all kind of thermal spray technologies and their applications can be found in "Thermal Spray Fundamentals" from Powder to Parts [13]. For PEM electrolyzers, the thermal sprayed coatings need to be highly conductive and thus required the absence of oxygen during the spraying process in order to prevent the formation of semi conductive titanium oxides. In this context, the use of a vacuum chamber is necessary for producing high purity titanium coatings.

2.2. Coated low-cost bipolar plates

In conventional PEM water electrolysis systems, the bipolar plates are often one of the most expensive components accounting for 51% of the stack cost [11, 15–17]. This is due to the high amount of expansive materials such as titanium and the complicated production of this component [18]. The bipolar plates need to meet the following requirements [15, 19–21]:

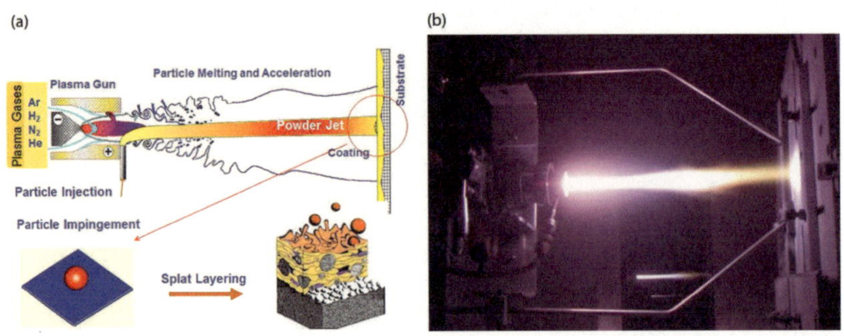

Figure 2. (a) Scheme of VPS spraying technology and (b) photo of the plasma torch [13].

1. Separation of the gases on the anode and the cathode side
2. Current transfer from the positively charged anode to the negatively charged cathode side
3. Optimal water distribution over the active area toward the anode-side electrode structure
4. Transport of the produced gases

Given these specifications, the materials used for manufacturing conventional bipolar plates must have the following properties [15, 20]:

1. Good electrical and thermal conductivity for very high currents and small contact resistances
2. Optimized flow field for the distribution and inflow of water to the electrochemically active surface of the electrodes and at the same time optimized removal of the produced gases
3. Cathode side: resistance to hydrogen embrittlement and oxidation
4. Anode side: corrosion resistance against oxygen in acidic media and voltages up to 2 V

Rarely, the anode and cathode sides of the bipolar plates are physically two separate plates. That is why the corrosive operating conditions of the anode side and the contact to the acidic membrane determine the material selection [16, 21]. In fact, bipolar plates consist almost exclusively of titanium. The material price of pure titanium (grades 1–2) [6] is comparatively high and the processing is difficult due to its brittleness [18, 22]. Often, a flow field must be chemically etched, resulting in an increase of production cost. Under the aforementioned conditions, the formation of a passive and poorly electrically conductive oxidation layer on titanium is undesirable [23, 24]. Furthermore, the formation of unstable hydroxides at the cathode should be prevented. Both phenomena lead to ohmic losses due to increase of contact resistance and drop in efficiency [25]. These negative effects can be prevented by an additional coating or surface modifications with noble metals [18–28]. Platinum is for this purpose the state of the art material.

In commercial PEM electrolyzers, the anode side requires an anticorrosive layer when using alternative bulk materials such as stainless steel, copper or aluminum to substitute titanium. The corrosion of non-noble metals would release dissolved ions, which poison the MEA, resulting into a strong degradation [28, 29]. On the other hand, these are easy-to-process materials can be coated with corrosion resistant materials through various coating processes. Thin film coatings of Au [30], TiN [31], TiN/C [32], TaN [33] and SnO_2: F [34] have already been investigated for protecting metallic bipolar plates for PEM fuel cells. All these materials and coatings can hardly withstand the strongly oxidative conditions of the cathode side of the fuel cells and would do so even less under PEM electrolysis conditions. In this regard, a thick film coating is necessary and vacuum plasma spraying technology (VPS, 50 mbar) was chosen to produce protective coatings of Ti for stainless steel (ss, 316 L) BPPs. Titanium particles of 45 μm were completely melted by achieving a plasma enthalpy of 21.3 MJ kg^{-1} and deposited on preheated stainless steel BPPs. The number of sweeps of the torch over the substrate defines the thickness of the coating. Post-treatment of capillary sealing with an epoxy resin allows for closing possible holes end imperfections and guarantees the complete separation of the stainless steel from the acidic environment of the electrolyzer.

Figure 3a shows a scanning electron microscopy (SEM) image of the Ti coating produced by VPS on a stainless steel BPP (92 E Stack, Hydrogenics). The coating covers the whole surface of the flow field channels as well as the edges and contact area. **Figure 3b** presents a magnified image to the coating where the small holes and imperfections are observable. The rough surface of the coating is polished after the sealing process to provide a conductive and uniform contact area to the porous transport layer. Nevertheless, an additional thin film coating is necessary to prevent the oxidation of the titanium coating and reduce the contact resistance for lower overpotentials, resulting in higher efficiencies.

Platinum is a common material used in the industry as anti-passivation coating for titanium. It has an outstanding conductive behavior but is highly expensive and rare and the substitution of this material by cheaper ones is a pressing challenge. The alternative material for Pt as a surface modification needs to fulfill all the previously mentioned requirements, having better electric properties than TiO_2 but lower cost than Pt. Different valve metals, such as Nb, V, Zr, W, Ti, Ta, Hf, etc. can be considered for this purpose while Nb is the most promising one given its low cost (almost three orders of magnitude lower price) and abundance on the earth crust. Thin films of Pt and Nb were deposited on the thermal sprayed coatings of Ti by magnetron sputtering physical vapor deposition (PVD). The measurement of the interfacial contact resistance (ICR) of Nb/Ti and Pt/Ti coatings on stainless steel (**Figure 4**), Nb/Ti/ss and Pt/Ti/ss, depicts the significantly improved electric behavior of Nb/Ti compared to uncoated Ti, yet not comparable to Pt/Ti. The results suggest that Nb coatings by thermal spraying can be sprayed directly on stainless steel to function as both, corrosion protective layer and surface modification for better conductivity.

Cross-section images of Ti-coated samples with surface modifications of Nb can be seen in **Figure 5a** and **b**. **Figure 5c** presents a stainless steel sample coated Nb coating without the intermediate layer of Ti. The image highlights the marked differences between the two substrates for the Nb film which influence the morphology and the adherence properties. The Nb

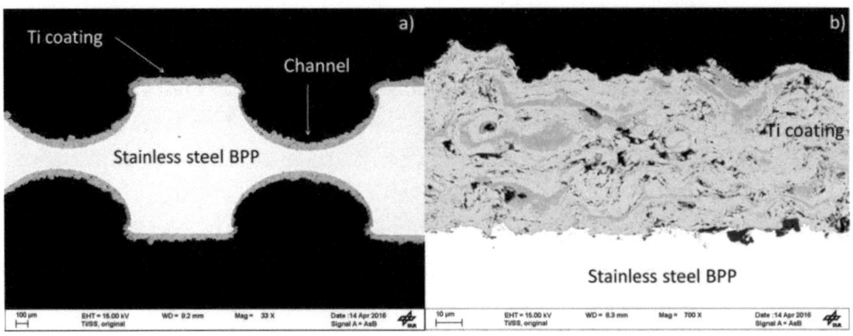

Figure 3. (a) SEM image of coated stainless steel BPP (92E Stack, Hydrogenics) with flow field channels and (b) cutaway of the Ti coating.

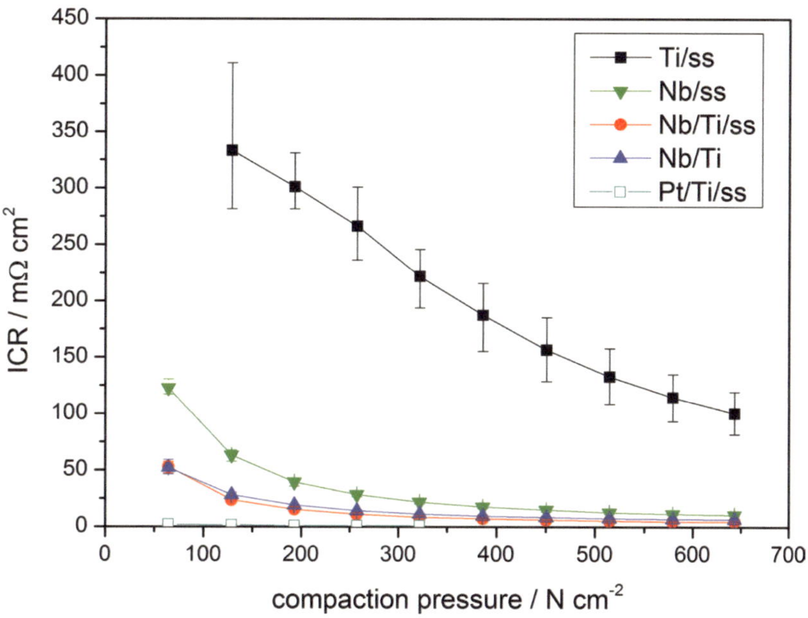

Figure 4. Interfacial contact resistance (ICR) of Nb-coated titanium, Ti-coated stainless steel without surface modification as well as with surface modification by Pt and Nb respectively [35, 36].

coating has preferred orientated crystals in (110) with the size between 10 and 20 nm. While some peaks for Ti can be observed in the X-ray diffraction (XRD) presented in **Figure 5d**, no reflections of stainless steel components can be detected.

Corrosion tests in simulated PEM electrolysis environment account for the high stability of the coatings. Initially, potentiodynamic measurements were performed at low potential sweep rates to determine the corrosion current (i_{corr}) and potential (E_{corr}) of the pristine samples. Afterwards an accelerated stress test (AST) for 6 h at 2 V vs. Reversible Hydrogen Electrode (RHE) in O_2-saturated 0.5 M H_2SO_4 was performed, and the resulting current transient represents the stability against corrosion and dissolution of the substrate. Finally, a second potentiodynamic measurement records the material changes. As can be seen in **Figure 6a**, all Ti-coated samples show almost exactly the same behavior than bulk titanium foil (blue curve). However, one can see that the thin coating of Nb by magnetron sputtering PVD is not sufficient to protect the stainless steel against corrosion (inset **Figure 6a**) and the sample experiences pitting corrosions. The current transient for Nb on stainless steel is constantly increasing at a current density level of almost three orders of magnitude higher than the other samples which passivate after approx. 1 h.

The potentiodynamic characterization of the materials is able to provide a detailed overview of the corrosion processes. Before the AST (**Figure 6a**), the E_{corr} for all samples is approx. 0 V in which the cathodic reaction of hydrogen evolution takes place. While all protected samples show a wide passivation region up to 2 V vs. RHE, the Nb/ss specimen displays a trans-passive region at potentials higher than 1.2 V vs. RHE where Fe and Cr oxides dissolve. These oxides as well as molecular O_2 are reduced during the second characterization protocol (**Figure 6c**)

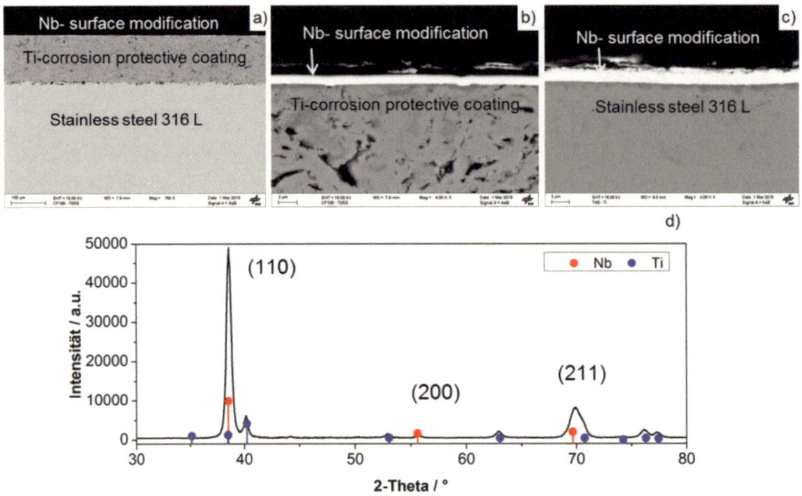

Figure 5. (a) and (b) SEM images of stainless steel samples with Nb/Ti coating. (c) Nb film directly deposited on stainless steel and (d) the XRD spectra of Nb/Ti/ss [35].

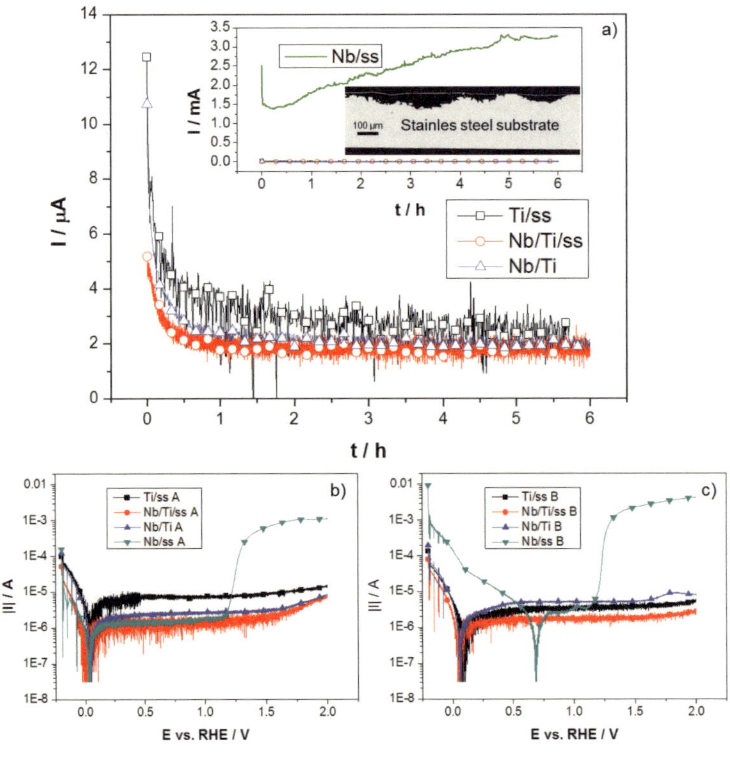

Figure 6. Galvanostatic measurements (AST) of the samples Ti/ss, Nb/Ti/ss, Nb/Ti and insert for Nb/ss with adjusted scale in (a) as well as potentiodynamic material characterization before (b) and after (c) the AST [35].

in a cathodic reaction, shifting the corrosion potential of Nb/ss to 0.7 V vs. RHE. From these results, one can conclude that the layer of Ti is able to protect the stainless steel substrates against corrosion, while a much thinner PVD coating is not sufficient for this purpose.

The use of stainless steel as bulk material of BPPs can reduce the material cost. Additional cost reduction can be achieved since stainless steel is easier to machine than titanium. The price of the Ti coatings can be estimated to 30.8 USD m^{-2} if industrial production rates can be achieved. This is in sum less than the half for coated stainless steel BPPs than for massive Ti BPPs. The use of Nb instead of Pt as surface modification can additionally reduce the material costs for BPPs. However, the superior electrical properties of Nb vs. Ti observed for the ICR measurements need to be demonstrated in real conditions of PEM electrolyzers. A rainbow stack was assembled to investigate the impact of surface material on anode and cathode side. The coatings were tested and compared to commercially coated-titanium BPPs. The cells are separated in two categories: category 1 corresponds to cells with precious metal surface modification and category 2 to cells with valve metals (Ti or Nb) on the surface on the cathode side, see **Table 1**.

The stainless steel BPPs were coated on both sides with titanium to avoid any possible galvanic corrosion of the cathode surface when in contact with the MEA during standby periods. However, titanium, in contrast to stainless steel, is not susceptible against hydrogen embrittlement. Therefore, a protective coating or surface treatment on the cathode side for the massive titanium BPPs, as well as for Ti-coated stainless steel ones, is required. This assumption is supported with the results presented in **Figure 7**, which correspond to a long-term performance of the rainbow stack. The lower the cell potential (E_{cell}) is, the higher the efficiency of the PEM electrolyzers. Cells 4, 5 and 8, which have either Ti or Nb/Ti coatings on the cathode side, show from the very beginning of the test, significantly worse performance compared with those having Pt. The high E_{cell} can be explained by hydrogen embrittlement of the valve metals. After 500 h, the electrolyzer was switched off for several hours, and oxygen was able to diffuse to the cathode side passivating even further the material. The consequence is a rapid potential increase in the next 500 h of the test. The commercial cells and those coated with Pt

Cell	BPP anode	BPP cathode	Category
1	Commercially coated Ti	Commercially coated Ti	1
2	Commercially coated Ti	Commercially coated Ti	1
3	Ti/ss	Commercially coated Ti	1
4	Ti/ss	Ti/ss	2
5	Pt/Ti/ss	Ti/ss	2
6	Pt/Ti/ss	Pt/Ti/ss	1
7	Nb/Ti/ss	Pt/Ti/ss	1
8	Nb/Ti/ss	Nb/Ti/ss	2

Table 1. Rainbow stack configuration: stainless steel BPPs (92E Stack, Hydrogenics) with the corrosion resistant Ti coating: Ti/ss. BPPs with additional surface modification of Pt and Nb, Pt/Ti/ss and Nb/Ti/ss, respectively. E300 and E400 MEAs (Greenerity) were used [35, 36].

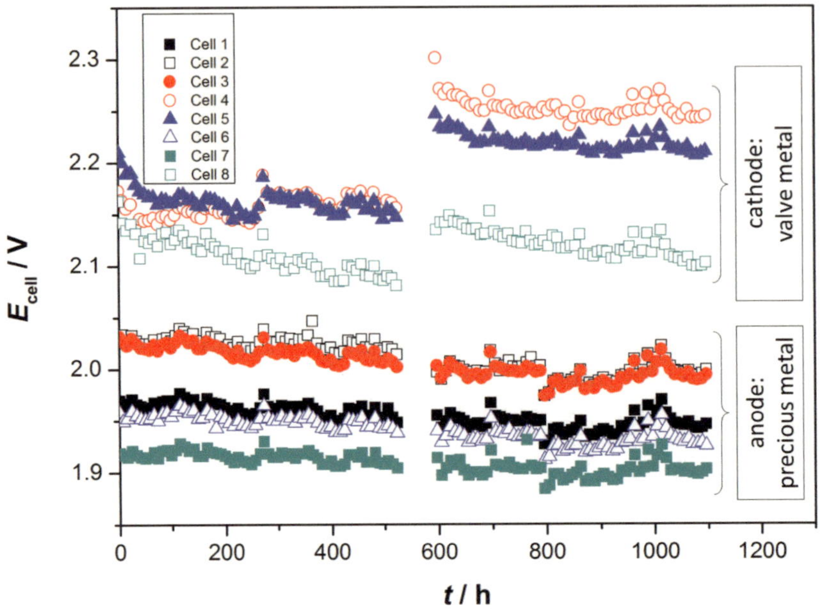

Figure 7. Durability test for 1000 h at constant current density of 1 A cm^{-2}.

on the cathode side show in general a much higher performance. Here, it is also possible to distinguish the positive impact of the Pt and Nb surface modification for the anode side. Cell 3 has higher performance as cell 6 and 7 which have surface modifications. Those cells show also a stable behavior during the entire testing protocol.

The post-mortem analysis of the BPPs proves the protective behavior of the thick film coating of Ti even after 1000 h in real PEM electrolysis conditions. The SEM images of the BPPs show in every case the protective effect of the Ti coating. Differences can be seen for the thin film surface modification. For the anode side, Pt and Nb are still in well adhered to the titanium coating (**Figure 8b**) and **d**), but for the cathode side, only Pt is still in contact to Ti. **Figure 8c** shows clearly that the Nb thin film coating is delaminated allowing for the oxidation of the titanium beneath the Nb.

Summarizing, Ti coating produced by VPS is able to protect the stainless steel BPPs which decrease significantly the material and fabrication cost of these components. In principle, it should be enough to coat the anode side since stainless steel is highly resistant to hydrogen embrittlement. Surface modifications with Nb for reducing the interfacial contact resistance can substitute the expensive Pt without compromising performance and durability. Finally, Nb could be sprayed by VPS directly on the stainless steel substrate thus avoiding the use of PVD techniques, which are expensive and have low deposition rates.

2.3. Production of porous transport layer via VPS

The porous transport layer (PTL), current collector (CC) or gas diffusion layer (GDL) is another key component of commercially available PEM electrolyzers. It assumes the following tasks:

1. Transport of electrons from the catalytic layer to the bipolar plate
2. Transport of the gases produced at the electrode
3. Distributes of the water toward the electrode

Thus, an optimum of the electrical conductivity and the contact resistance as well as the mass transport of the gases and water in the, respectively, opposite direction needs to be met. In addition, the PTL on the anode side requires a large number of additional properties similar to the BPPs:

1. Corrosion resistance for voltages up to 2 V
2. Suitable pore size and porosity [37]
3. Mechanical stability especially against hydrogen embrittlement from cathode crossover
4. Sufficiently low thickness for an optimized removal of the resulting gases
5. Low contact resistance to the adjacent components: Anode catalyst layer and BPP
6. Possibility of up-scaling for industrial production of PEM electrolyzers in the MW to GW range

On the cathode side, the state of the art is carbon paper, which is predominantly used in PEM fuel cell technology. Fine stainless steel fibers, sintered plates and meshes can also be used. Given the moderate conditions on the cathode side, the research concentrates currently to the anode side current collector [38]. On the oxygen side, the environment is more corrosive for these interconnectors because of the pure oxygen atmosphere, low pH and voltages of up to 2 V. For these reasons, corrosion resistant meshes, fibers, foams to sintered structures of titanium or platinum are used [37–44], **Figure 9**.

The electrical resistance of the current collector depends on the porosity and the contact resistance of the bipolar plate with the current collector and electrode. With decreasing pore size and porosity, the electrical resistance of the current collector decreases [37, 40]. Optimum

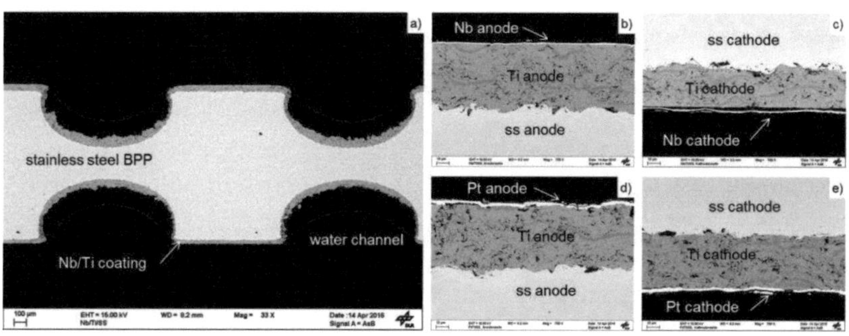

Figure 8. Post-mortem SEM analysis after 1000 h in use as electrolyzer BPP. (a) overview, (b) Nb/Ti/ss anode, (c) Nb/Ti/ss cathode, (d) Pt/Ti/ss anode, (e) Pt/Ti/ss cathode [35].

Figure 9. Samples of porous transport layers: a) sintered titanium and below carbon GDL; SEM images of b) sintered titanium, c) titanium felt, d) carbon GDL and e) mesh. Images courtesy of Fraunhofer ISE.

porosity of 20–50% and pore sizes between 10 and 13 μm are determined with regard to electrical resistance vs. mass transport issues [37]. With the use of sintered materials, the porosity and pore size can be adjusted by the appropriate choice of the titanium particles, but the possibilities for upscaling are very limited [45]. On the one hand, meshes, fibers and foams are less limited in terms of thickness and shape [46], and they are also cheaper to manufacture [18], but the control of the pore size and porosity is more complex.

The interfacial contact resistance (ICR) from the current collector to the electrode is a decisive factor for the reduction of the ohmic resistance and thus for the efficiency of a PEM cell [40]. The ICR can be greatly reduced by the use of precious metal coatings such as platinum [24, 36, 47, 48]. In PEM fuel cell technology, the problem of water management and contact resistance has been solved by the use of a microporous layer (MPL) [49] which is applied to the gas diffusion layer. The use of micrometer-sized, conductive and hydrophobic particles improves both the contact resistance [49] and the water transport [50]. More than 149 refereed publications have been published since 2002 on this topic in connection with the fuel cell but just one for PEM electrolysis (ISI Web of Knowledge) [45].

Thereby it is possible to produce coatings with similar properties as the MPLs by VPS out of titanium or other stable valve metals such as Nb. The benefit is to have a controlled contact layer to the electrode, optimized contact surface, pore sizes, porosity, etc. The performance of electrolyzer is typically characterized by a polarization curve. The potential is recorded at different current densities. **Figure 10** displays the performance of an electrolysis cell with and without MPL. The inset shows a cross section SEM image of the MPL deposited on the sintered Ti current collector. It can be seen that the polarization curve of the cell with MPL has a lower slope than the one without MPL. The slope of the current potential curve is related to a ohmic drop caused by a decrease of contact resistance. At high current densities and consequently high gas production rates, the cell with MPL shows no mass transport limitation meaning an optimized mass transport behavior, while the cell with standard PTL increases rapidly the potential and a significant loss in efficiency is produced.

The mass transport properties of the MPL can be further improved by VPS coating technology, and it can be applied on Ti meshes and all other mechanical stable structures. Furthermore, by

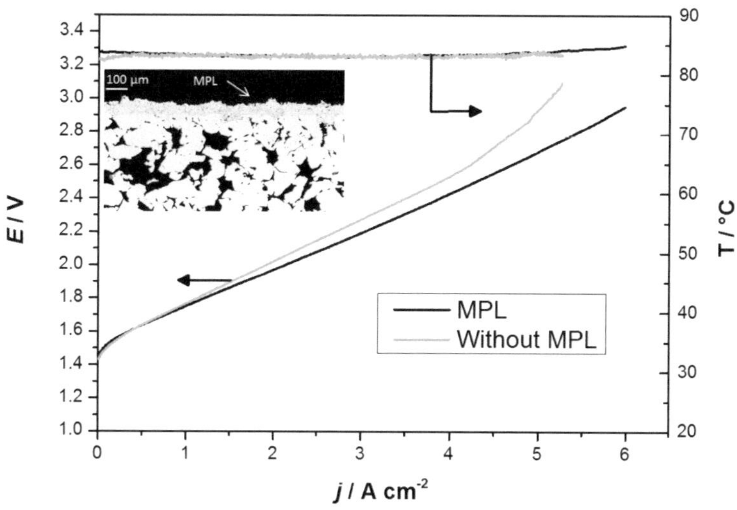

Figure 10. Polarization curves up to 6 A cm^{-2} of PEM electrolyzer cells with and without MPL. The inset shows an SEM image of the MPL deposited on the sintered Ti current collector [44].

adjusting the spraying parameters, it is possible to produce such porous transport layer without limitation of surface area by reducing the plasma enthalpy one is able to control the state of particle melting and produce highly porous structures. **Figure 11** shows some examples of different porous structures produced with different Ti particle sizes, sweep numbers and plasma enthalpies. **Figure 11a** depicts the production of a free-standing thin, highly porous structure similar to the commercial sintered titanium. By use of smaller particles, a mixture of sintered titanium and MPL can be produced which is mechanically stable and able to stand

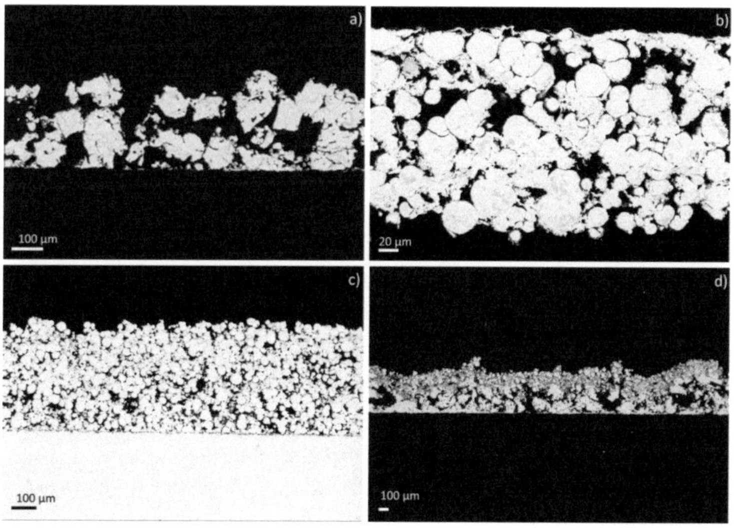

Figure 11. SEM cross-section images of (a) free-standing porous layer of 125 µm particles, (b) free-standing porous layer of 45 µm particles, (c) porous layer of 45 µm particles on BPP and (d) multi-functional layer of different particle sizes.

alone as an independent porous transport layer (**Figure 11b**) or can be sprayed directly on the BPP (**Figure 11c**). With VPS technology, thickness, porosity, pore size of the current collectors can be finely tuned to determine important parameters such as capillary pressure, bubble point and tortuosity.

3. Summary

Proton exchange membrane (PEM) electrolysis is a very promising technology for a sustainable hydrogen production and a comprehensive use of renewable energy. PEM electrolyzers are efficient, have high energy densities, and are flexible enough to play an important role for the integration of fluctuating renewables by power grid stabilizing effects. For a large-scale commercialization, the technology needs to become more economically by production cost reduction. Vacuum plasma spraying (VPS) technology is a promising tool to solve pressing question about material substitution. By controlling the spraying parameters, as well highly dense and protective layer can be produced as well porous structures for use as porous transport layers. It handles thereby with the most expensive components and allows the use of stainless steel as basis material for BPPs, which can be protected by VPS coatings. This saves material and fabrication costs. By the use of Nb, it is potentially possible to produce a protective but simultaneously highly conductive coating which may reduce the BPPs cost significantly. By reducing the plasma enthalpy, the powder material can be just partly melted to produce low cost and highly efficient components for use as porous transport layer or any other function such as filter material. The possibilities are endless, and VPS is therefore a promising technology, which can dominate the future PEM electrolysis landscape.

Author details

Philipp Lettenmeier[1], Aldo S. Gago[1*] and K. Andreas Friedrich[1,2]

*Address all correspondence to: aldo.gago@dlr.de

1 Institute of Engineering Thermodynamics, German Aerospace Center (DLR), Pfaffenwaldring, Stuttgart, Germany

2 Institute for Energy Storage, University of Stuttgart, Keplerstraße, Stuttgart, Germany

References

[1] United Nations, Adoption of the Paris Agreement, 2015. https://unfccc.int/resource/docs/2015/cop21/eng/l09r01.pdf

[2] United Nations, Paris Agreement—Entry into Force, 2016. http://unfccc.int/paris_agreement/items/9444.php

[3] Rogelj J, Luderer G, Pietzcker RC, Kriegler E, Schaeffer M, Krey V, et al. Energy system transformations for limiting end-of-century warming to below 1.5°C. Nature Climate Change. 2015;**5**:519-527. DOI: 10.1038/nclimate2572

[4] Royal Society of Chemistry, Enterprise and Electrolysis. 2003. http://www.rsc.org/chemistryworld/Issues/2003/August/electrolysis.asp [Accessed 10-March-2016]

[5] Russell JH, Nuttall LJ, Fickett AP. Hydrogen generation by solid polymer electrolyte water electrolysis. American Chemical Society, Division of Fuel Chemistry Preprints of Papers. 1973;**18:**pp. 24-33

[6] Carmo M, Fritz DL, Mergel J, Stolten D. A comprehensive review on PEM water electrolysis. International journal of Hydrogen Energy. 2013;**38**:4901-4934. DOI: 10.1016/j.ijhydene.2013.01.151

[7] Friedrich KA. PlanDelyKad: Study on Large Scale Water Electrolysis and Hydrogen Storage (in German). Berlin: German Federal Ministry for Economic Affairs and Energy (BMWi); 2015, (n.d.)

[8] Eichman J, Harrison KW, Peters M. Novel Electrolyzer Applications: Providing more than just Hydrogen, National Renewable Energy Laboratory, 2014. https://books.google.com/books?id=X3q3rQEACAAJ&pgis=1 [Accessed 17-December-2015]

[9] Sutherland E, II. 4 Low-Cost Large-Scale PEM Electrolysis for Renewable Energy Storage, 2013. pp. 22-25. https://www.hydrogen.energy.gov/pdfs/progress13/ii_a_4_ayers_2013.pdf

[10] Lettenmeier P, Wang R, Abouatallah R, Helmly S, Morawietz T, Hiesgen R, et al. Durable membrane electrode assemblies for proton Exchange membrane electrolyzer systems operating at high current densities, Electrochimica Acta. 2016;**210**:502-511 DOI: 10.1016/j.electacta.2016.04.164

[11] Bertuccioli L, Chan A, Hart D, Lehner F, Madden B, Eleanor Standen: Fuel cells and hydrogen joint undertaking development of water electrolysis in the European Union. 2014. http://www.fch.europa.eu/sites/default/files/study%20electrolyser_0-Logos_0_0.pdf

[12] Pasumarthi V, Chen Y, Bakshi SR, Agarwal A. Reaction synthesis of Ti3SiC2 phase in plasma sprayed coating. Journal of Alloys and Compounds. 2009;**484**:113-117. DOI: 10.1016/j.jallcom.2009.04.079

[13] Fauchais PL, Heberlein JVR, Boulos M. Thermal Spray Fundamentals—From Powder to Part, (n.d.). http://www.springer.com/de/book/9780387283197 [Accessed 24-October-2016]

[14] Fauchais P. Understanding plasma spraying. Journal of Physics D: Applied Physics. 2004;**37**:R86-R108. DOI: 10.1088/0022-3727/37/9/R02

[15] Hermann A, Chaudhuri T, Spagnol P. Bipolar plates for PEM fuel cells: A review. International Journal of Hydrogen Energy. 2005;**30**:1297-1302. DOI: 10.1016/j.ijhydene.2005.04.016

[16] Antunes RA, Oliveira MCL, Ett G, Ett V. Corrosion of metal bipolar plates for PEM fuel cells: A review. International Journal of Hydrogen Energy. 2010;**35**:3632-3647. DOI: 10.1016/j.ijhydene.2010.01.059

[17] Ayers KE, Capuano C, Anderson EB. Recent advances in cell cost and efficiency for PEM-based water electrolysis. ECS Transactions. 2012;**41**:15-22. DOI: 10.1149/1.3684798

[18] Kim HG, Kwa LK, Han W, Kwac LK, Han W. The performance and stability of a PEM electrolyzer using 3-D mesh hong. In: International Conference on Electrical Power and Energy Systems. Lecture Notes Information Technology; Vol.13, 2012. pp. 373-379

[19] Wang JT, Wang WW, Wang C, Mao ZQ. Corrosion behavior of three bipolar plate materials in simulated SPE water electrolysis environment. International Journal of Hydrogen Energy. 2012;**37**:12069-12073. DOI: 10.1016/j.ijhydene.2012.04.146

[20] Lee SJ, Huang CH, Chen YP. Investigation of PVD coating on corrosion resistance of metallic bipolar plates in PEM fuel cell. Journal of Materials Processing Technology. 2003;**140**:688-693. DOI: 10.1016/S0924-0136(03)00743-X

[21] André J, Antoni L, Petit JP. Corrosion resistance of stainless steel bipolar plates in a PEFC environment: A comprehensive study. International Journal of Hydrogen Energy. 2010;**35**:3684-3697. DOI: 10.1016/j.ijhydene.2010.01.062

[22] Jin CK, Jeong MG, Kang CG. Fabrication of titanium bipolar plates by rubber forming and performance of single cell using TiN-coated titanium bipolar plates. International Journal of Hydrogen Energy. 2014;**39**:21480-21488. DOI: 10.1016/j.ijhydene.2014.03.013

[23] Jung HY, Huang SY, Ganesan P, Popov BN. Performance of gold-coated titanium bipolar plates in unitized regenerative fuel cell operation. Journal of Power Sources. 2009;**194**:972-975. DOI: 10.1016/j.jpowsour.2009.06.030

[24] Jung HY, Huang SY, Popov BN. High-durability titanium bipolar plate modified by electrochemical deposition of platinum for unitized regenerative fuel cell (URFC). Journal of Power Sources. 2010;**195**:1950-1956. DOI: 10.1016/j.jpowsour.2009.10.002

[25] Wang SH, Peng J, Lui WB. Surface modification and development of titanium bipolar plates for PEM fuel cells. Journal of Power Sources. 2006;**160**:485-489. DOI: 10.1016/j.jpowsour.2006.01.020

[26] Dihrab SS, Sopian K, Alghoul MA, Sulaiman MY. Review of the membrane and bipolar plates materials for conventional and unitized regenerative fuel cells. Renewable and Sustainable Energy. 2009;**13**:1663-1668

[27] Hodgson DR, May B, Adcock PL, Davies DP. New lightweight bipolar plate system for polymer electrolyte membrane fuel cells. Journal of Power Sources. 2001;**96**:233-235

[28] Wang X, Zhang L, Li G, Zhang G, Shao Z, Yi B. The influence of Ferric ion contamination on the solid polymer electrolyte water electrolysis performance. Electrochimica Acta. 2015;**158**:253-257

[29] Sun S, Shao Z, Yu H, Li G, Yi B. Investigations on degradation of the long-term proton exchange membrane water electrolysis stack. Journal of Power Sources. 2014;**267**:515-520. DOi: 10.1016/j.jpowsour.2014.05.117

[30] Kumar A, Ricketts M, Hirano S. Ex situ evaluation of nanometer range gold coating on stainless steel substrate for automotive polymer electrolyte membrane fuel cell bipolar plate. Journal of Power Sources. 2010;**195**:1401-1407. DOI: 10.1016/j.jpowsour.2009.09.022

[31] Wang Y, Northwood DO. Effect of substrate material on the corrosion of TiN-coated stainless steels in simulated anode and cathode environments of proton exchange membrane fuel cells. Journal of Power Sources. 2009;**191**:483-488. DOI: 10.1016/j.jpowsour.2009.02.029

[32] Sun H, Cooke K, Eitzinger G, Hamilton P, Pollet B. Development of PVD coatings for PEMFC metallic bipolar plates. Thin Solid Films. 2013;**528**:199-204. DOI: 10.1016/j.tsf.2012.10.094

[33] Choe C, Choi H, Hong W, Lee JJ. Tantalum nitride coated AISI 316L as bipolar plate for polymer electrolyte membrane fuel cell. International Journal of Hydrogen Energy. 2012;**37**:405-411. DOI: 10.1016/j.ijhydene.2011.09.060

[34] Wang H, Turner JA, SnO_2: F coated ferritic stainless steels for PEM fuel cell bipolar plates. Journal of Power Sources. 2007;**170**:387-394. DOI: 10.1016/j.jpowsour.2007.04.028

[35] Lettenmeier P, Wang R, Abouatallah R, Saruhan B, Freitag O, Gazdzicki P, Gago AS, Friedrich KA. Low-cost and durable bipolar plates for proton exchange membrane electrolyzers. Scientific Reports. 2017;**7**:44035; DOI: 10.1038/srep44035. in press

[36] Lettenmeier P, Wang R, Abouatallah R, Burggraf F, Gago AS, Friedrich KA. Coated stainless steel bipolar plates for proton exchange membrane electrolyzers. Journal of the Electrochemical Society. 2016;**163**:F3119-F3124. DOI: 10.1149/2.0141611jes

[37] Grigoriev SA, Millet P, Volobuev SA, Fateev VN. Optimization of porous current collectors for PEM water electrolysers. International Journal of Hydrogen Energy. 2009;**34**:4968-4973. DOI: 10.1016/j.ijhydene.2008.11.056

[38] Ito H, Maeda T, Nakano A, Hwang CM, Ishida M, Kato A, et al. Experimental study on porous current collectors of PEM electrolyzers. International Journal of Hydrogen Energy. 2012;**37**:7418-7428. DOI: 10.1016/j.ijhydene.2012.01.095

[39] Selamet OF, Pasaogullari U, Spernjak D, Hussey DS, Jacobson DL, Mat MD. Two-phase flow in a proton exchange membrane electrolyzer visualized in situ by simultaneous neutron radiography and optical imaging. International Journal of Hydrogen Energy. 2013;**38**:5823-5835. DOI: 10.1016/j.ijhydene.2013.02.087

[40] Ito H, Maeda T, Nakano A, Kato A, Yoshida T. Influence of pore structural properties of current collectors on the performance of proton exchange membrane electrolyzer. Electrochimica Acta. 2013;**100**:242-248. DOI: 10.1016/j.electacta.2012.05.068

[41] Arbabi F, Kalantarian A, Abouatallah R, Wang R, Wallace JS. Bazylak A. Feasibility study of using microfluidic platforms for visualizing bubble flows in electrolyzer gas diffusion layers. Journal of Power Sources. 2014;**258**:142-149. DOI: 10.1016/j.jpowsour.2014.02.042

[42] Siracusano S, Di Blasi A, Baglio V, Brunaccini G, Briguglio N, Stassi A, et al. Optimization of components and assembling in a PEM electrolyzer stack, International journal of Hydrogen Energy. 2011;**36**:3333-3339. DOI: 10.1016/j.ijhydene.2010.12.044

[43] Grigoriev SA, Dzhus KA, Bessarabov DG, Millet P. Failure of PEM water electrolysis cells: Case study involving anode dissolution and membrane thinning, International journal of Hydrogen Energy. 2014;**39**:20440-20446. DOI: 10.1016/j.ijhydene.2014.05.043

[44] Lettenmeier P, Kolb S, Burggraf F, Gago AS, Friedrich KA. Towards developing a backing layer for proton exchange membrane electrolyzers, Journal of Power Sources. 2016;**311**:153-158. DOI: 10.1016/j.jpowsour.2016.01.100

[45] Grigoriev SA, Dzhus KA, Bessarabov DG, Millet P. Failure of PEM water electrolysis cells: Case study involving anode dissolution and membrane thinning, International Journal of Hydrogen Energy. 2014;**39**:20440-20446. http://dx.doi.org/10.1016/j.ijhydene.2014.05.043

[46] Hwang CM, Ishida M, Ito H, Maeda T, Nakano A, Kato A, et al. Effect of titanium powder loading in gas diffusion layer of a polymer electrolyte unitized reversible fuel cell, Journal of Power Sources. 2012;**202**:108-113. DOI: 10.1016/j.jpowsour.2011.11.041

[47] Gago AS, Ansar SA, Saruhan B, Schulz U, Lettenmeier P, Cañas NA, et al. Protective coatings on stainless steel bipolar plates for proton exchange membrane (PEM) electrolysers. Journal of Power Sources. 2016;**307**:815-825. DOI: 10.1016/j.jpowsour.2015.12.071

[48] Rakousky C, Reimer U, Wippermann K, Carmo M, Lueke W, Stolten D. An analysis of degradation phenomena in polymer electrolyte membrane water electrolysis. Journal of Power Sources. 2016;**326**:120-128. DOI: 10.1016/j.jpowsour.2016.06.082

[49] Park S, Lee JW, Popov BN. Effect of carbon loading in microporous layer on PEM fuel cell performance. Journal of Power Sources. 2006;**163**:357-363. DOI: 10.1016/j.jpowsour.2006.09.020

[50] Weber AZ, Newman J. Effects of Microporous Layers in Polymer Electrolyte Fuel Cells. Journal of the Electrochemical Society. 2005;**152**:A677. DOI: 10.1149/1.1861194

Chapter 5

Hydrophobic Coatings for Corrosion Control of Aluminum Heat Exchangers

Alba Covelo, Carmina Menchaca, Miriam Flores,
Pilar Rodríguez-Rojas, Miguel Hernandez-Gallegos,
Esteban Martinez Meza,
Rebecca Jaimes-Ramírez and Jorge Uruchurtu

Additional information is available at the end of the chapter

http://dx.doi.org/10.5772/67676

Abstract

The production of thin films (nanocoatings) is a technological field with many applications to elaborate materials with new properties to be used as corrosion protection of traditional metals. Hydrophobicity is an example of such properties. In this chapter, an example of two hydrophobic corrosion coatings for possible use over aluminum heat exchanger geothermal power plants is discussed. Material substrate preparation, synthesis of hydrophobic sol-gel nanocoating, characterization, and electrochemical evaluation as a function of time of immersion, which are compared to another commercial fluorinated compound, are presented. Good corrosion protection was found for both hydrophobic coatings for possible application in geothermal heat exchangers.

Keywords: hydrophobic, coatings, corrosion, aluminum, geothermal heat exchangers

1. Introduction

The main problem of heat exchanger operation in contact with highly corrosive geothermal fluids is the deposition of fluid salts within heat exchanger tube walls, giving rise to corrosion problems. In combination with fluid pH and aggressive ions concentration such as chlorides, they reduce the lifetime operation of such equipment. Due to these facts and including mass transfer considerations, affordable protection of heat exchangers is a must from the effect of corrosive geothermal fluid. Various materials do exist and are used in heat exchangers in the geothermal energy production industries such as steel, copper alloys, and aluminum, depending on cost-benefit decision making. Nevertheless, proposals do exist to coat these materials with polymeric-coating systems giving dubious practical results in the past. A possible alternative is the use of hydrophobic coatings, to increase the long-term aluminum corrosion resistance under severe natural saline conditions encountered in geothermal fluid fields.

1.1. Nanocoatings

The production of thin films (nano coatings) is a technological field with many applications to elaborate materials with new properties as well as protection of traditional metals, nowadays and in the foreseeable future. Different materials and coatings are combined to produce hybrid or composite materials with special characteristics and properties [1].

The coating thickness range is from tens of nanometers (nm) up to some microns (μm) and in most *grand* scale applications, mono-layered coatings are used, but fabricating oxide-oxide, oxide-metal or other variable metals oxides combination multi-layers, properties may be improved. Therefore, film thickness control is critical [2, 3].

Coating can be classified due to their manufacturing material interactions, substrate nature, and according to its specific barrier applications in thermal, corrosion and oxidation, wear and diffusion types [4]. Coating nano-technology concentrates in obtaining thinner films with the same or enhanced protection when compared to conventional coating technologies.

1.2. Sol-gel coatings

Within the recent deposition technique alternatives, the sol-gel route technology is promising. In the last two decades, there is an increased widespread use of nanotechnology coatings through sol-gel methods, formed by way of organic-inorganic components, eliminating highly damaging and toxic solvents to the environment and the human itself. These coatings protect the metal surface from corrosion by different corrosion protection mechanisms [5].

Sol-gel synthesis route represents an alternative to generate components or compounds with controlled size, which was difficult to obtain in the past by conventional preparation techniques. This uniform procedure process results not only in particles agglutination but also in a range of products such as fibers, monolithic structures, thin films, or coatings in a wide variety of hybrid or composite materials. The adequate selection of precursor and reactants allows obtaining materials with tailor-made–designed properties, useful in a wide variety of technological applications such as optics, electronics, biology, medicine, etc. [6]. In general, the sol-gel route is the formation of a three-dimensional oxide resulting from hydrolysis and condensation reactions of the molecules from precursors present in a liquid medium under relatively low temperatures and no physical and/or chemical reactions with the substrates, these being advantageous for practical applications.

1.3. Hydrophobic coatings

Hydrophobic coatings reject an aqueous dissolution or electrolyte. These characteristics may be accomplished by means of encapsulated functionalized species present in the coating or by changing the medium composition. Also, if the more external or outer-layer morphology or structure of the system is changed or modified, a hydro- or superhydrophobic system is promoted [7–10]. The strategies or methodologies contemplating a superficial change through encapsulation systems promised advantageous success; nevertheless, few developments really achieved an effective corrosion protection, since the effect decreases as a function of elapsed time of surface (coating) contact electrolyte [11–13].

To obtain a hydrophobic outcome, a "*Lotus leave*" effect must be generated, which is when the surface attains a roughness level that effectively repels any type of aqueous dissolution. To simulate this effect on coating surfaces, various types of species, such as potassium stearate or calcium hydroxide, porous silica or synthetic urea formaldehyde capsules, styrene-based copolymers such as methyl-methacrylate, were encapsulated and incorporated into coating systems, among others [7–14].

Sol-gel is a versatile method to produce super-hydrophobic surfaces. The literature reports [15] successive hydrolysis and condensation-polymerization reactions using ammonia as a catalyzer over aluminum substrate, generating films with high degree of hydrophobicity (contact angles ≈150°). Good corrosion resistance for short periods of immersion was obtained. Incorporation of porous silica particles doped with 3-amino-propyl triethoxysilane compounds as hydrophobic agent generates transparent sol-gel coatings [16]. Also, sol-gel modifies the polymeric structure with diverse functionalized agents such as fluorinated compounds [17] and/or organic-inorganic precursors [15], or nano-fibers or nanotubes incorporation [18] promoting Lotus leaves-like nanostructures increasing corrosion resistance.

1.4. Aluminum and alloys

Aluminum and alloys are widely used in the industrial, architectural, and marine environments due to technical and economic reasons, being the most extended metallic material used, after steel. Nevertheless, they are very reactive and the need for extra protection against corrosion is necessary for certain industrial applications, especially in chloride-containing environments. To achieve this goal, coating and surface modification technologies were speedily developed in recent times related to chemistry, electrochemistry, metallurgy, and other disciplines. Aluminum and alloy surface modification generally consists of surface roughness generation (etching), anodic oxidation, hybrid or composite coatings, etc. The chemically modified surface samples present super-hydrophobicity, increasing corrosion resistance of aluminum and its alloys. This result is attributed to the combined effect of nano/microstructures formed over the surface and low-energy surface material. Chemical etching improves the hydro- or superhydrophobic properties of aluminum and alloys.

1.5. Corrosion in geothermal heat exchangers

In principle, the fluid extracted from the geothermal well is taken to a separator where the mixture, water vapor from the geothermal fluid, is separated. Vapor is sent to the turbine coupled to a generator, where mechanic energy is transformed into electrical energy. The turbine exit is coupled to a condenser helping to increase the cycle efficiency. Finally, the geothermal fluid is reinjected to the well to help recharging.

The main function of heat exchangers metallic tubes or plates is the heat transfer from a heated flow to the feeding water, and for that reason parameters like material thermal conductivity and thickness should be considered. Metallic corrosion promotes thinning of the tubes or plate walls, causing huge economic losses due to operation failures and plant shutdowns as well as malfunctioning of heat exchange processes. Incrustation is the undesirable material accumulation over the metallic elements, calcium carbonates being the most common precipitation, although silica and metallic sulfurous compounds are also common [19]. Both problems present a decrease in equipment performance and efficiency, since overall heat-exchange coefficient diminishes gradually promoting some component failure.

Geothermal environments present different composition with respect to the type of resources available, low or high enthalpy [19]. However, the composition solely does not depend on this; in low-enthalpy environments, the main forms of corrosion are generalized (overall) and localized (certain areas) corrosion. Localized corrosion is the most detrimental damage, since it cannot be predicted and is difficult to follow up its evolution and it can produce major damages. This type of corrosion includes galvanic, pitting, crevice; in general, three types of material incrustations may be encountered: silica and silicates, carbonates, and sulfates and sulfites [20]. Silica is in the form of amorphous silica, carbonate incrustations are in the form of low magnesium calcite and sulfates, while crystallized sulfite in many phases predominates as Pb, Zn, Fe, and Cu [20].

The pH control in geothermal water is a crucial factor to control corrosion and silica incrustations; to control the system, HCl or NaOH is added to the geothermal energy source, reactants being expensive. The alternative is to use protective coatings to help diminish incrustations within the tube walls.

The main problem during geothermal plant heat exchanger operation in contact with geothermal fluids is deposition of fluid salts over the tube walls originating multiple corrosion problems which in combination with the fluid pH and aggressive ions concentration (mainly chlorides) notably reduce the useful life of heat exchangers. Therefore, and taking into account heat transfer considerations, it is necessary to protect the heat exchanger metallic elements from the highly corrosive fluid. Different materials are used as heat exchangers in the geothermal industry like steel, copper alloys, or aluminum and proposals are being made to coat the metallic elements with different polymeric systems or schemes [21, 22].

2. Coatings

Examples of hydrophobic coatings were obtained from sol-gel, or dip-coating methods, over aluminum substrate. The hybrid-coating systems used were based on silane polymer solutions or compounds. Evaluation and characterization were done using electrochemical techniques and characterization through scanning electron microscope (SEM), water-drop contact angle, and thermal conductivity measurements.

2.1. Electrochemical etching

Commercial aluminum (see **Table 1**) rods were cut in cylinders 2 cm height to expose approximately 1 cm^2 area to the corrosive electrolytes, which after their electrical connection by means of a copper wire were encapsulated in an epoxy resin at room temperature. Before experiments, the electrodes' surface was prepared abrading with 600 grade emery paper and rinsed with distilled water, and after cleaning, the electrodes were degreased with acetone and finally dried in warm air flow. The purpose is twofold: to render the metal surface ready for coating application improving the coating adhesion and second to improve corrosion resistance of the metal surface. The electrochemical arrangement was a three-electrode cell using a graphite rod as a counter-electrode and a silver/silver chloride as a reference electrode.

The aluminum etching is as follows: the electrode was immersed in NaOH pH 11 solution for about 3 min. Afterwards, the electrodes were removed and cleaned again with de-ionized water, dried and immersed again, this time in a HCl pH 3 solution, promoting anodic dissolution, polarizing up to −100 $mV_{(Ag/AgCl2)}$ potential and left constant at this potential for 30 min. The presence of chloride ions promotes the formation of micro-/nanopits or pores over the metal surface. The samples were ultrasonically cleaned with de-ionized water, dried, and ready for coating application.

Figure 1 presents SEM micrographs (cross section) where micro/nanostructure can be observed over the aluminum surface, after electrochemical etching and anodic dissolution. Rough and convex irregular and attacked areas and pitted surface are clearly observed. These structures may be favoring the capacity to trap air bubbles supporting hydrophobicity with an appropriate surface roughness. The structure is irregular and disordered and micro/nanoparticles and pores can be observed.

In the presence of chloride ions, the surface structure promotes the appearance of a larger quantity of pits and nanopores at high potentials (above -100 mV). This is presented in general and a detailed view is seen in **Figure 2**, where a multitude of pits and pores over the anodized surface can be observed. The anodized compact film surface diminishes corrosive anions attacking the aluminum substrate [23].

2.2. Coatings

Then, the as-prepared samples were immersed in 5mM perfluoro cthyl trioxi sylane {$CF_3(CF_2)_5(CH_2)_2$-$Si(OCH_2CH_3)_3$} (PTES) solution obtained from Sigma Aldrich. The solution prepared was a mixture of de-ionized water and ethanol (volume ratio 1:1) at 60°C for 2 h [24]. The coating was applied through two-cycle dip coating technique. The first one was immersion for 1 min and then removed and was allowed to dry, and then a second 30-s immersion was done to seal the pores present on the coated sample.

Material composition (%)						
Aluminum	Mg	Fe	Mn	Zn	Cu	Si
	0.014	0.83	0.03	0.9	0.04	0.28

Table 1. Aluminum substrate composition.

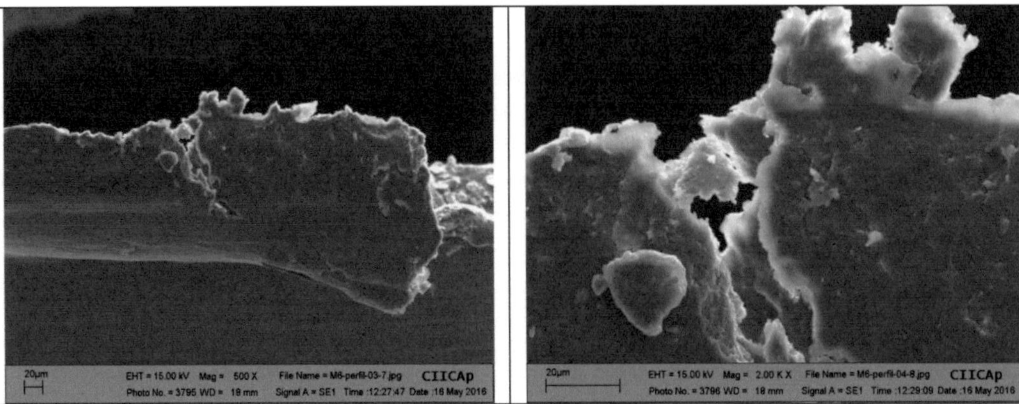

General view of Al structure after etching.	Detailed view showing micro/nano structures.

Figure 1. Micrograph after electrochemical aluminum etching in acid media.

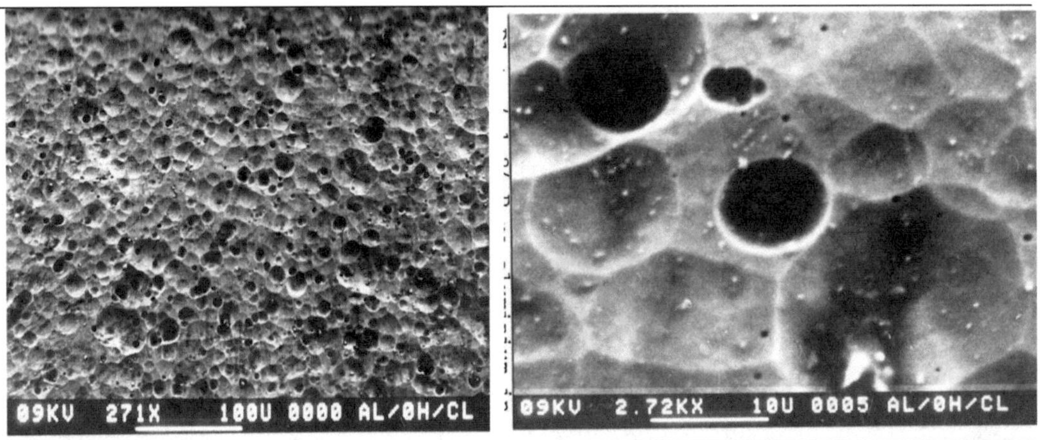

General view of the anodized aluminium in the presence of chlorides	Detailed view of micro/nano pores formed during anodic oxidation

Figure 2. SEM micrograph showing pores over aluminum in acid media in the presence of chloride ions.

Another coating was prepared via sol-gel synthesis, combining one inorganic and another organic compound. The first one consists of a combination of zirconium tetra-n-propoxide with ethyl aceto-acetate using nitric acid pH 5, as a catalyzer for the reaction. The organic part was prepared mixing 3-glycyloxypropyltrimethoxysilane (GPTMS) with 2-propanol (SA nomenclature) also in the presence of HNO_3. During the sol-gel synthesis, a commercial fluoride compound tetra deca-flouro hexane (TDFH) *Chemguard* from Sigma Aldrich (HA nomenclature) was incorporated in different weight proportions (10, 20, and 30%). The sol-gel coating was deposited using the spin-coating technique, at 2800 rpm for 28 s.

http://dx.doi.org/10.5772/67676

2.2.1. Coating thickness

After substrate preparation and coating application, film thickness measurements were taken using a coating thickness tester. The average values obtained from five measurements are presented in **Table 2**.

The highest thickness obtained and observed was the PTES coating, while the sol-gel coatings were 10 times thinner. From these coatings, the higher thickness obtained was the SA coating and the CH 30%, with the highest fluoride compound percentage incorporated. As an example, **Figure 3** presents the general and detailed view (cross section) of the aluminum PTES-coated sample, showing good adhesion conditions over the metal substrate.

2.2.2. Hydrophobicity

For each sample, the contact angle (right and left) was measured and the average was obtained. When a surface shows a contact angle greater than 90°, it represents a hydrophobic surface. The PTES- and CH 20%-coated samples clearly present hydrophobicity (see **Table 2**). The image is presented in **Figure 4**.

Coating	Thickness (μm)	Contact angle (degrees)
Blank	1.99	66.7
PTES	24.2	115
SA10%	2.27	70.3
CH 10%	2.02	82.6
CH 20%	2.02	118.9
CH 30%	2.36	95.8

Table 2. Coating thickness measurements.

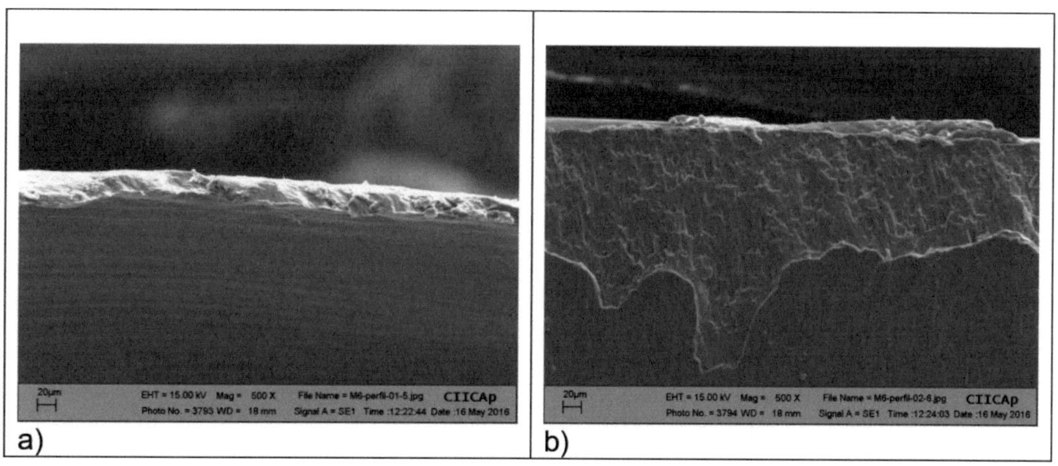

Figure 3. SEM micrograph showing PTES coating: (a) general and (b) detailed view.

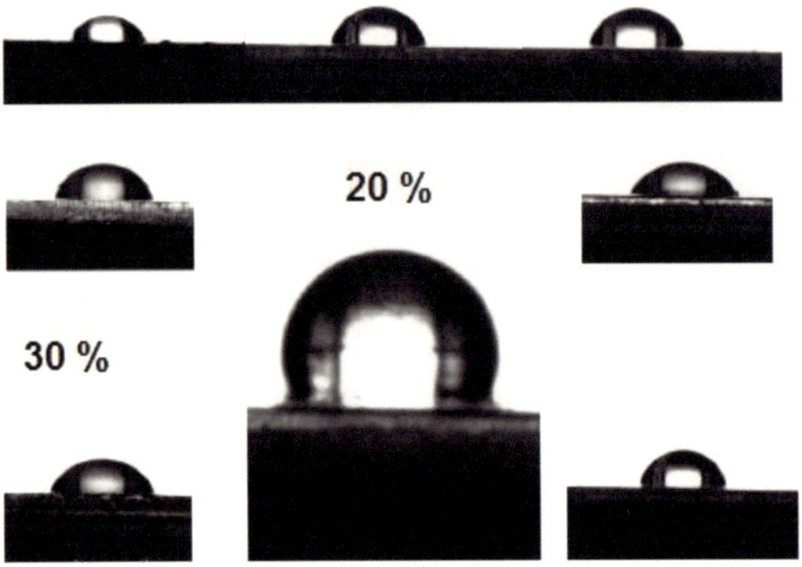

Figure 4. Contact angle of coating.

2.2.3. Thermal conductivity

One method to evaluate the thermal conductivity (heat transfer) of polymer coatings, which is important to use in heat exchangers, is the heat plate method to measure the temperature distribution of thermal conductivity. This is according to ISO 8302:1991 standard [25]. Its principle consists in generating a unidirectional heat flow through the samples, as flat plates

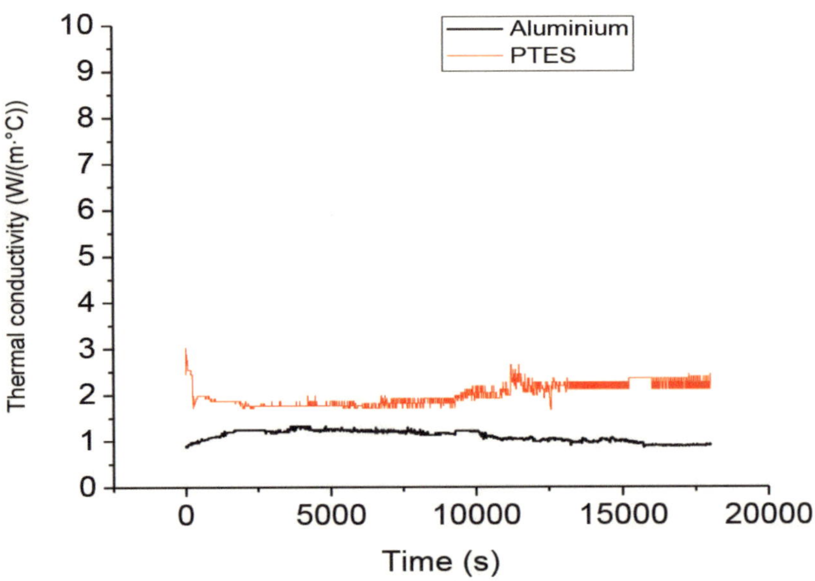

Figure 5. Thermal conductivity of bare aluminum and PTES-coated sample.

http://dx.doi.org/10.5772/67676

conducting heat. As an example, **Figure 5** presents the thermal conductivity as a function of time of PTES coating over the aluminum substrate and compared with the blank (bare metal) sample. The coating presents favorable thermal conductivity conditions when compared with aluminum.

3. Electrochemical evaluation

To evaluate the performance and efficiency of corrosion protection coatings, it is common to use electrochemical methods, electrochemical impedance spectroscopy (EIS) being widely applied. The working electrode was immersed in the test solution similar to a geothermal fluid containing 3580 ppm Na and 6693 Cl until a steady-state open-circuit potential (E_{oc}) was measured. Electrochemical impedance spectroscopy measurements as a function of time were made using an ACM potentiostat, carried out at corrosion potential E_{oc}, by using a small sinusoidal signal with an amplitude of ±10 mV, in a frequency interval of 50 mHz to 20 KHz recording six points per decade. The EIS results were represented using Nyquist and Bode plots, and the effect of time of exposure of the aluminum and coated samples to the corrosion media was also evaluated. Also, electrochemical noise measurements (ENMs) were taken with a second electrode at the tip of a platinum wire and the reference electrode.

The parameters for electrochemical noise were a second Pt tip electrode 1 or 0.5 s per sample obtaining 1024 measurements. The electrolyte was (3580 ppm Na and 6693 ppm Cl) similar to a geothermal fluid (see **Figure 6**).

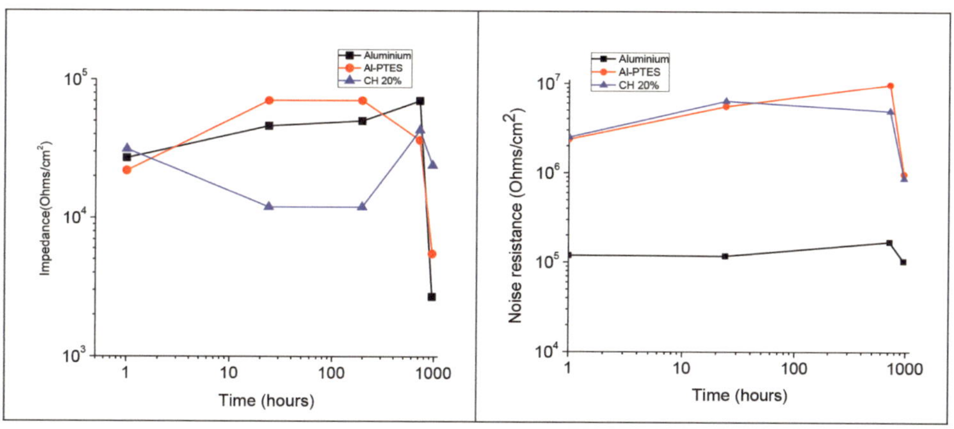

Figure 6. Electrochemical measurements as a function of time: EIS charge transfer resistance and electrochemical noise resistance.

From the electrochemical results obtained EIS charge transfer resistance and electrochemical noise resistance, a similar pattern is revealed. After 720-h immersion, a decrease was observed for both techniques, suggesting an increase in the corrosion rate for the coatings tested. A decrease of one order of magnitude was observed [26].

4. Conclusions

An example of hydrophobic coatings for geothermal heat exchangers was presented. The coating characterization was obtained and coating evaluation as a function of time was presented. Good coating behavior was obtained for both hydrophobic coatings tested.

Acknowledgements

Rebecca Jaimes Ramírez acknowledges Programa de Becas Posdoctorales from DGAPA, Universidad Nacional Autónoma de México 2016-2017, and J. Uruchurtu thanks CONACYT for financial support during sabbatical stay.

Author details

Alba Covelo[1], Carmina Menchaca[2], Miriam Flores[2], Pilar Rodríguez-Rojas[2], Miguel Hernandez-Gallegos[1], Esteban Martinez Meza[2], Rebecca Jaimes-Ramírez[1] and Jorge Uruchurtu[2]*

*Address all correspondence to: juch25@uaem.mx

1 UNAM-CENISA, Ciudad de Mexico, Mexico

2 UAEM-IICBA, Cuernavaca, Mexico

References

[1] Durán A., 2000. Sol-gel produced coatings. Ciencia e Ingeniería de la Superficie de los Materiales Metálicos. Textos Universitarios C.S.I.C., Madrid. pp. 59-70.

[2] Pulker H. K., 1984. Coatings on Glass. Edit. Elsevier. Amsterdam. pp. 18-23.

[3] Dislich, H. 1983. Glassy and crystalline systems from gels: chemical basis and technical applications. J. Non-Cryst. Sol., vol. 57, pp. 371-388.

[4] Niemi K. 1985. Proc. Int. Conf. Thermal Spraying: Current Status and Future Trends, ITSC'95, Japan. pp. 687.

[5] Wang D., Bierwagen G. P., 2009. Sol-gel coatings on metals for corrosion protection. Prog. Org. Coat., vol. 64. pp. 327-328.

[6] Handbook of Sol-Gel Science and Technology. 2005. Processing, Characterization and Applications. Edited by Sumiosakka, Kluwer Academic Publishers. pp. 618-643.

[7] Wang Q., Li J., Zhang C., Qu X., Liu, Z J., Yang Z., 2010. Regenerative super-hydrophobic coating from microcapsules, J. Mater. Chem. vol. 20. pp. 3211-3215.

[8] Deng X., Mammen L., Zhao Y., Lellig P., Müllen K., Li C., Butt H. J., Vollmer D., 2011. Transparent, thermally stable and mechanically robust superhydrophobic surfaces made from porous silica capsules, Adv. Mater. vol. 23. pp. 2962-2965.

[9] Zhang L., D'Acunzi M., Kappl M., Auernhammer G.K., Vollmer D., van Kats M., van Blaaderen A., 2009. Hollow silica spheres: synthesis and mechanical properties Langmuir, vol. 25. pp. 2711-2717.

[10] Huang Y.F., Huang C., Zhong Y.L., Yi S.P., 2013. Preparing superhydrophobic surfaces with very low contact angle hysteresis. Surf. Eng. vol. 29. pp. 633-636.

[11] Steele A., Bayer I., Yeong Y.H., De Combarieu G., Lakeman C., Deglaire P., Loth E., 2011. Adhesion strength and superhydrophobicity in polyurethane/organo clay nanocomposites, NSTI-Nanotech Conf. vol. 1. pp. 399-417.

[12] Sethi S., Dhinojwala A., 2009. Super hydrophobic conductive carbon nano tube coatings for steel, Langmuir, vol. 25. pp. 4311-4315.

[13] Patel C.J., Dighe A., 2007. Novel isocyanate-free self-curable cathodically depositable epoxy coatings: Influence of epoxy groups on coating properties, Prog. Org. Coat. vol. 60. pp. 219-223.

[14] García S.J., Fischer H.R., White P.A., Mardel J., González-García Y., Mol J. M. C., Hughes A.E., 2011. Self-healing anticorrosive organic coating based on an encapsulated water reactive silyl ester: synthesis and proof of concept, Prog. Org. Coat vol. 70, pp. 142-149.

[15] Mu M., Liang J., Zhou X., Xiao Q., 2013. One-step preparation of TiO 2/MoS 2 composite coating on Ti6Al4V alloy by plasma electrolytic oxidation and its tribological properties, Surf. Coat. Technol. vol. 214, pp. 124-130.

[16] Xu L., He J., 2012. Fabrication of highly transparent superhydrophobic coatings from hollow silica nanoparticles, Langmuir, vol. 28. pp. 7512-7518.

[17] Brassard J.D., Sarkar D.K., Perron J., 2011. Synthesis of monodisperse fluorinated silica nanoparticles and their superhydrophobic thin films, ACS Appl. Mater. Interfaces, vol. 3 pp. 3583-3588.

[18] ldarkhanova F.I., Mironova G.A., Bogoslovsky K.G., Menshikov V.V., Bykov E.D., 2012. Methods of nanostructuring and hydrophobization of paint coating surface are studied. It is found that the volume modification of the paint with carbon nanotubes and nanofibers affords nanostructuring Prot. Met. Phys. Chem. Surf. vol. 48. pp. 796-802.

[19] Kristmannsdóttir H., 1989. Types of scaling occurring by geothermal utilization in Iceland. Geothermics, vol. 18, pp. 183-190,

[20] Corsi R., 1986. Scaling and corrosion in geothermal equipment: problems and preventive measures. Geothermics, vol. 15, pp. 839-856.

[21] Lund J., Tonya L. Boyd M., 2016. Direct utilization of geothermal energy 2015 worldwide review, Geothermics, vol. 15, pp. 66-93.

[22] Barbier E., 1997. Nature and technology of geothermal energy: a review. Renew. Sustain. Energy Rev. vol. 1, pp. 1-69.

[23] Hoar T.P., Mears D. C., Rothwell G. P., 1965. The relationships between anodic passivity, brightening and pitting, Corros. Sci., vol. 5, pp. 279-289.

[24] Wu Y., Zhao W., Wang W., Sui W., 2016. Fabricating binary anti-corrosion structures containing super hydrophobic surfaces and sturdy barrier layers for Al alloys. RCS Adv, vol. 6, pp. 5100-5110.

[25] ISO 8302:1991 Standard, Thermal insulation—determination of steady-state thermal resistance and related properties – Guarded hot plate apparatus.

[26] Hernández-Escampa M., Rodríguez-Acuña F., Millán-Cruz F., Rodríguez-Rojas P., Hernández-Gallegos M., Covelo A., Menchaca-Campos C., Uruchurtu J., 2013. Nylon/graphene oxide electrospun composite coating, J. Polym. Sci. ID 751056, pp. 1-9.

Section 3

Fire Action on Materials

Chapter 6

Fire Retardant Coatings

Thirumal Mariappan

Additional information is available at the end of the chapter

http://dx.doi.org/10.5772/67675

Abstract

Fire retardant coatings are often required to protect a wide range of products of both flammable and nonflammable against fire. It is an oldest, most efficient, and easiest method to apply any surface without modifying the intrinsic properties of materials. Moreover, the initial phase of fire always occurs on the surface by ignition, and hence, it is important to concentrate on the surface protection of a material. Being an organic nature of conventional surface coating will burn easily and generate smoke and toxic fumes, which may not be suitable for application where fire protection or fire prevention is required. Reaction-to-fire and/or resistance-to-fire are to be considered for assessing both flammable and non-flammable material by using fire retardant and fire resistant or fire protective coatings. The degree of fire retardation mainly depends on the coating thickness, substrates, and efficiency of formulations. This chapter explains briefly the fire retardation of wood by using fire retardant coatings.

Keywords: fire retardant, wood, intumescent coating, charring, nanocomposites, sol-gel process

1. Introduction

The devastating nature of fire creates havoc that result in great loss of both lives and assets. It can also cause serious human sufferings and financial losses. Thus, great attention has been paid to develop the effective fire protection methods in order to prevent any casualties and reducing economic fire damage to an acceptable level. Nowadays, because of rapid changes in building and developments in architectural technology have led to more complex structure of which using a strong and lightweight components, the wide range of materials in the form of organic or inorganic and synthetic or natural used in building are easily flammable or readily burning on exposure to high temperatures of fire. On the other hand, non-combustible material such as metals (except aluminum and magnesium) and concrete will not burn and support the combustion, or release the flammable vapors when subject to fire. However, it is important to consider these materials can withstand a fire for a specified period of time. In this regard, the strength and deformation of these materials deteriorate significantly at high temperature. As a result, the structural elements and assemblies may deform or even collapse when exposed to fire conditions. The allowable or extra time in a fire situation depends largely on the anticipated temperature development of the fire, which depends on the type and amount of combustible materials present and the ventilation condition [1–3]. Although protection of materials against fire by the use of coatings for indefinite periods is impossible, it can delay the spread of fire or keep a structure intact against fire, thereby allowing sufficient time for safety measures to be taken. This chapter covers with a discussion of the fire retardation of wood products that can be available in scientific literatures.

2. Fire retardant vs. fire resistant

There is always misconception about inhibition of fire by using coating of either fire retardant or fire resistance. **Figure 1** shows the development of fire in a compartment, which distinguishes the usage of coating in a particular stage of fire. For instance, the material properties play an important role in prior to flashover, which will be controlled by fire retardant coatings (ignition, flame spread, release of heat, smoke, and gases), whereas fire resistant coatings are mainly predominated in protection of structures that occur after flashover, that is, phase of fully developed fire. Post-flashover fires (temperature above 600°C) are considered as hazardous fires for structures [4]. Fire retardant coatings are mainly involved for reaction-to-fire to retard or inhibit the combustion of flammable materials (wood, foam, textile fabrics, electric cables, and fiber reinforced composites) whereas fire resistant or fire protective coating for resistance-to-fire to protect the non-flammable materials. Different test parameters, such as oxygen index (OI), flame spread rate, ignition time, heat intensity, smoke generation, and

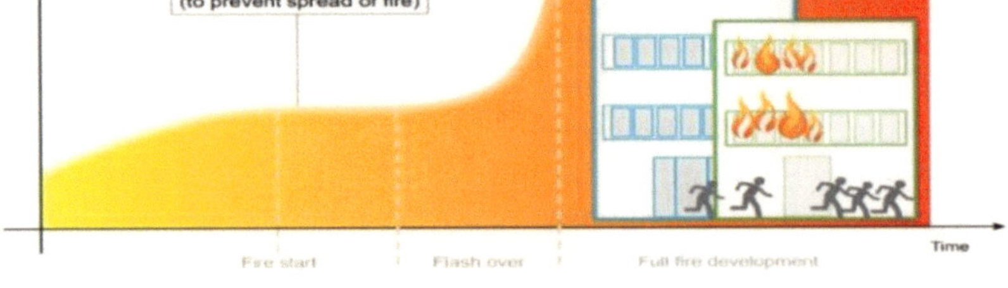

Figure 1. Phases of fire development in an enclosure.

http://dx.doi.org/10.5772/67675

release of toxic gases, are to be considered for assessing the flammable materials. On the other hand, thermal insulation, integrity, and load-bearing capacity are paramount important for the fire resistance of non-flammable materials. Depending on the types of material and intended application, specific fire performance properties of building materials are to be tested by use of different test methods. The relevant test standards for both coatings are different that also depend on the region of countries.

3. Fire retardant coatings

In general, conventional organic surface coating is easily ignitable, melts, drips, and it may cause severe injury and damage to the substrates in the event of fire. Therefore, coatings that are designed to formulate should not contribute a significant amount of fuel to the fire and, at the same time, limit the flame spread and smoke development. Fire retardant coating is one of the easiest, oldest, and most efficient ways to protect the materials against fire. This approach does not cause chemical modification of the substrate, but rather the formation of a protective layer which alters the heat flux to the substrate and can inhibit its thermal degradation, ignition, or combustion [5]. The ideal fire retardant coatings should have minimum flame spread, negligible or low release of smoke and/or toxic gases, be easy to apply, show good wear resistance, adhere to the underlying substrate and offer low cost. Typically, they are based on chlorinated alkyds or brominated epoxy resin and filled aluminum hydroxide or a combination of chlorinated paraffin and antimony oxide system. However, the wide range of flame retardants in the form of reactive or additive and different classes (halogen, phosphorus, nitrogen, minerals, oxides, intumescent, and nanofillers) are available commercially that are incorporated into coating formulations to inhibit or retard the burning of materials. The selection of flame retardants can be tailor-made for a specific polymer binder in coating formulations. Fire retardant coatings look like architectural paints and mainly available in solvent form and are applied by conventional methods, brush, roller, and spray.

Fire retardant coatings can be classified into two groups: non-intumescent and intumescent coatings [6]. Non-intumescent coatings are basically decorative, architectural coatings that contain flame retardant additives designed to reduce the rate of flame spread and smoke development of combustible substrates. They are rated as Class A, B or C based on the ability to not contribute to fire and smoke. Rate of flame spread depends on both substrate and thickness of the film. On the other hand, intumescent coatings swell under the influence of heat to form a multicellular charred layer, which acts as an insulating barrier and slows heat and mass transfer between the condensed and vapor phases. This intumesced char can increase to up to 50 times the original thickness of the applied coating. Two types of fire retardant coating either pigmented/colored or clear, transparent varnish are available on the market that are designed for use on different materials and that respond very differently when exposed to fire. They are mainly used in construction, transport, wall and ceiling linings, and other areas require products to satisfy the requirement of classes.

The burning of surface coating is believed to proceed by a free radical mechanism where both $^{\cdot}H$ and $^{\cdot}OH$ radicals are chain carriers and take part in a number of reactions in the flame zone. The function of fire retardant coatings is to protect the substrate that depends on the mode of operation of flame retardants [7]. For example, the coatings formulated with halogenated compounds are more effective in gas phase, and they act in the flame zone by forming a blanket of halogen vapor that interferes with the propagation of the flame by interrupting the generation of highly reactive free radicals, thus helping flame extinguishments. However, the release of toxic and corrosive gases from halogenated compounds while burning is ecologically unsafe. The action of phosphorus in coatings varies with the type of flame retardants and the polymer binders. They mostly operate in condensed phase to form a protective char layer that acts as physical barrier to heat transfer from the flame to the substrate and to diffusion of gases. The vapor phase action is also found to be effective in phosphorus flame retardants, which are capable of controlling the high energy radicals in the flame. Intumescent flame retardant is a combination of an acid source, a char former and a gas source, and sometimes, it is available in a single compound including all three functions. The mechanism of intumescent coatings is to undergo an endothermic decomposition reaction at an elevated temperature that causes the coating to swell and form into a highly porous, thick, and thermally stable char layer that has a very low thermal conductivity (heat insulation). Other flame retardants such as metal oxides and hydroxides are operated by cooling with the release of water and by diluting or removing the flammable fuels and oxygen.

4. Fire retardancy of flammable materials

Typically, the flammable materials are easily combustible and rapidly growing in a fire (reaction-to-fire), in terms of the spread of fire or propagation of fire, up the stage when flashover occurs in a compartment. Flashover can occur quickly in seconds or slowly depending on the speed of fire growth rate. Wood is the most frequently used combustible products in addition with polymers (plastics and rubbers), foams, textiles, cables, and fire reinforced composites. Three different kinds of methods, such as clear or transparent varnish paints, pigmented intumescent reactive coatings, and surface impregnations are utilized in wood to enable the restriction of growth and/or spread of fire. The European fire classification for reaction-to-fire is based on fire growth rate index (FIGRA), which indicates the time to reach flashover in the standardized reference test as per BS EN 14390. Other test methods, the single burning item (SBI) test (BS EN 13823; 2002), radiant panel test for flooring (EN ISO 9239-1; 2002) and either the small flame test (BS EN 11925-2; 2002) or the bomb calorimeter (BS EN ISO 1716), are also relevant to the fire retardant system in Euro-class [8]. Two standards, such as fire propagation test (BS476, part 6:1989) and surface spread of flame test (BS476, part 7:1987), were applicable to the fire test methods in the UK for flammable materials. The BS476, part 6 test method is intended to provide a comparative measure of the contribution to the growth of fire of a product. The test result is expressed as fire propagation index (I) and three sub-indices, i1, i2, and i3. The higher the fire propagation index means the greater the growth of fire. On the other hand, the BS476, part 7 measures the lateral spread of flame along the surface of a specimen, which is mounted at right angles to a high intensity radiation panel. The extent and rate of flame spread of specimen are used to determine the classification, which can range from Class

1 (the best) down to Class 4. The ASTM E 84, 2010 and ISO 5660 are most commonly used in USA and International standards for testing of flammable materials. The factors influencing the performance of coating systems are thickness, density, substrates, composites, and panel type. Each type of product shall be evaluated and representative specimens shall be tested to ensure that the effect of each variable parameter is considered.

5. Wood

Wood is one of the most versatile, sustainable, aesthetically pleasing, and environmentally benign materials. It can be classified into hard and soft wood, which can have different percentages of cellulose, hemicellulose, and lignin. Different types of wood products including solid wood-based panels (particleboard, hardboard, fiberboard, fir Douglas plywood), structural timbers, glued laminated timbers, cladding and wood floorings are widely used for structural purposes in building construction, flooring and furnishing materials that found in homes, schools, and offices around the world. The basic deficiencies of wood products are flammability, poor dimensional stability, and low resistance to micro-biological decay that must be addressed when used as a construction material. Because of easily flammable and contribute fuel to fires, wood is considered to poor construction materials. The flaming combustion of wood is mostly supported by cellulose [9].

The various categories of action are typically described in wood treated with flame retardants: (1) an acceleration of dehydration and carbonization that provides thermal insulation, (2) chemical modification of wood pyrolysis, (3) absorb the surrounding heat by endothermic reactions, (4) inhibition of the flaming combustion in the gas phase, and (5) increase the thermal conductivity of wood in order to dissipate the heat from the wood surface [10]. The fire retardation of wood is typically treated with flame retardant chemicals that are coated onto the surface of wood by painting, spraying or dipping methods and/or impregnated into the wood structure using vacuum pressure technique or plasma treatments [11], which inhibits ignition and do not contribute to the spread of flame. Three components of wood have quite different decomposition range; for instance, temperature from 200 to 260°C is for hemicellulose, temperature from 240 to 350°C is for cellulose, and temperature from 280 to 500°C is for lignin. When heated, wood undergoes degradation and combustion to produce volatile gases, tars (levoglucosan), and carbonaceous chars. The fire performance of wood-based products and test methods has been reviewed and studied extensively [12–22]. Traditional flame retardants, such as boron compounds, mineral acids, and inorganic salts, (monoammonium phosphate, diammonium phosphate, guanylurea phosphate, guanidine phosphate, ammonium polyphosphate, and melamine phosphate), may considerably improve the fire retardant properties of wood [23]. However, the use of boron and formaldehyde-based systems is likely to be declined in accordance with the growing awareness of environmental issues and consumer safety. In addition, inorganic salts may also affect the performance of wood in various ways by increasing hygroscopicity, reducing strength that leads to dimensional instability, wood degradation, corrosion of metal fasteners, adhesion problems, and increased abrasiveness. The flame retardants based on phosphorus, nitrogen, silicone, and char forming additives are still prominent solution to address the flammability and environmental issues [24].

Cone calorimeter is a most commonly used bench-scale method to evaluate the flammability of wood. Shi and Chew have investigated the carbon monoxide (CO) yield of six species of wood samples under different external heat fluxes and moisture content by spontaneous ignition in a cone calorimeter. Spontaneous ignition is a complex phenomenon that combustible materials are ignited by internal heating, without the spark plug. As compare to piloted ignition, process of spontaneous ignition is much closer to the development of real fire. Results observed that thickness of wood has little effect to peak CO release rate, but the time to peak is postponed with a higher thickness. The peak CO release rate decreases with a higher external heat flux, but the decrease is not obvious when heat flux increases from 50 to 75 kW/m^2. Average CO yield is inversely proportional to external heat flux, thickness, and density. They concluded that both flame and moisture can also reduce CO release rate because energy used for water evaporation increases with high moisture content [25]. The effect of variable heat flux and oxygen concentrations (20.9, 18, 16 and 15%) on ignition time and mass loss rate of wood was investigated to obtain the kinetic parameters, activation energy and frequency factor. It was found that with increasing the oxygen concentration, the mass loss rate was increased, but the ignition time, the activation energy and the frequency factor were decreased [26]. Critical heat flux for ignition has been calculated to be between 10 and 13 kW/m^2 for a range of wood products. Density, thickness and moisture content have a large influence on the material dependent properties [27].

6. Charring rate of wood

The charring rate, speed at which charring depth advance in the material when exposed to high temperature, is a critical parameter for flammability of wooden samples because it allows the determination of the size of the residual section of wood. It depends on wood species, density, moisture content, permeability, composition and direction of burning [28]. For example, the charring rate of flame retardant-treated wood is linearly proportional to the applied heat flux in cone calorimeter and inversely proportional to the density of wood. Charred wood is bounded by the transition between the pyrolysis layer, the zone where thermal degradation of wood and char formation is actually occurring and the char layer, a zone of cracked charcoal that has no relevant strength or stiffness properties. Charring depth is the distance between the outer surface of the original member and the position of the char line. The base of char layer is widely occurs between 280 and 300°C. Beikircher et al. determined the charring rate, mass loss and temperature development of Norway wood coated with transparent and colored intumescent coatings using cone calorimeter and ISO 834 furnace test. They found that the intumescent coatings reduce the charring rate significantly at all irradiance and cellulosic fire exposure condition (ISO 834 test curve) in comparison with the uncoated (REF) wood sample (**Figure 2**). Intumescent coatings can delay the onset of charring and reduce the charring rate of wood [29].

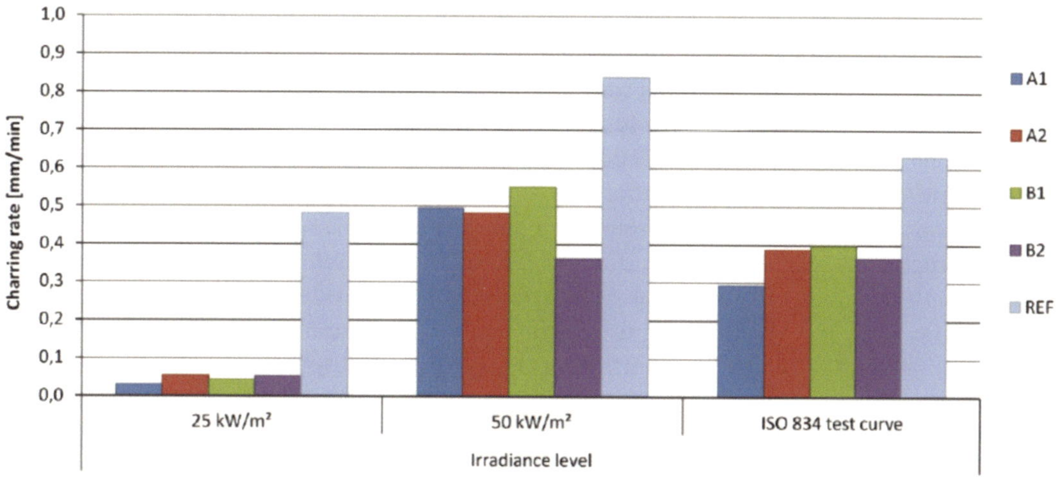

Figure 2. Charring rate of intumescent coated and uncoated wood exposed to different test conditions [29].

7. Boron-based fire retardant coatings

Boric acid (H_3BO_3) and borax ($Na_2B_4O_7.10H_2O$) are used as borates, which are water soluble and most commonly used flame retardants in wood products. Boron-containing formulations are also used to improve the service life of timbers, in terms of both increase in resistance to biological attacks and renders more resistant to burning. Atar and Keskin studied the flammability of varnish coated Uludag Fir wood and boron compound impregnation. The ASTM D1413-99 standard is used to impregnate the wood by vacuum technique using a mixture of boric acid and borax. The flammability of wood was characterized by ASTM E160-50. They found that boron impregnations showed decrease the flammability of wood as compared to varnish coated wood. It was suggested that the impregnation of wood with boron compounds before varnish coating can decrease the combustion temperature and provide security to wood structure. Further, they investigated the impact of boron compounds impregnation on combustion properties of the laminated veneer lumber obtained from European oak and Lombardy poplar woods. Because of the interaction of impregnation materials with wood structure, the lowest flame source combustion was observed in treated wood [30, 31].

Blasi et al. have investigated the pyrolysis products of fir wood impregnated with boric acid at heating temperatures of 377 and 527°C. It was found that the yields of char and water increase with boric acid concentration (below 2%), and the amounts of organic liquid products are reduced. The boric acid treatment lowers the activation energy and delayed the most important oxidation reaction of fir wood. The reaction temperature does not affect the pyrolysis product distribution but at lower temperature, the higher the char yields, water and non-flammable volatile products were observed [32]. Further, the effect of diammonium phosphate (DAP) and diammonium sulfate (DAS) treatment on the pyrolysis of wood was investigated. Two ammonium salts are widely used as flame retardants in wood substrate, and both are significantly alter the char reactivity. A lower activation energy and a higher reaction order are obtained for DAP-treated sample as compared to wood treated with DAS [33]. Both treatments produce an equal amount and composition of solid, liquid and gaseous products during decomposition. However, with increasing the concentration of salts and/or decreasing the heating temperature produce greater amount of char and water. It was concluded that DAP treatment showed better flame retardancy on the basis of the formation of higher yields of char, water and lower combustible flammable liquids. It also confirmed that the release of decomposed volatile products depends on the DAP concentration [34, 35]. The thermogravimetric analysis was carried out in air of wood and wood impregnated with DAP concentrations range up to 20% and heating rates between 5 and 20°C/min. Results showed a three step decompositions in sequence of wood decomposition, induction and char oxidation, and concluded that the estimated kinetic parameters are independent of the heating rate but vary with the DAP concentration. However, the activation energies of the various steps remain practically constant except for the decomposition of the cellulose component or the decomposition step, depending on the complexity of the mechanism [36].

The fire performance of Douglas fir wood was studied by both natural extractives and a mixture of boric acid and borax treatment. Dual treatments of wood with the natural extractives and borates were targeted to benefit from their potential cumulative protections, which are biological resistance and fire retardancy. It was observed that both treated wood specimens showed excellent fire retardant performance [37]. Tomark and Cavdar studied the effect of boron powder (BP), mixture of boric acid (BA) and borax (BX) and flame retardant agent (FA) based on liquid blend of limestone and silicone oil (SO) treatment on the oxygen index (OI) of Scots pine wood of bare and after leaching process. Leaching procedure was carried out to determine the permanent performance of the preservatives in wood. The oxygen index (OI) is the minimum percentage of oxygen required to continue flaming combustion of a sample under laboratory condition. Wood samples were initially vacuum treated with the preservative and then were subjected to leaching. **Figure 3** showed that the wood treated with flame retardants provided the best results, and moreover, leaching did not considerably change the

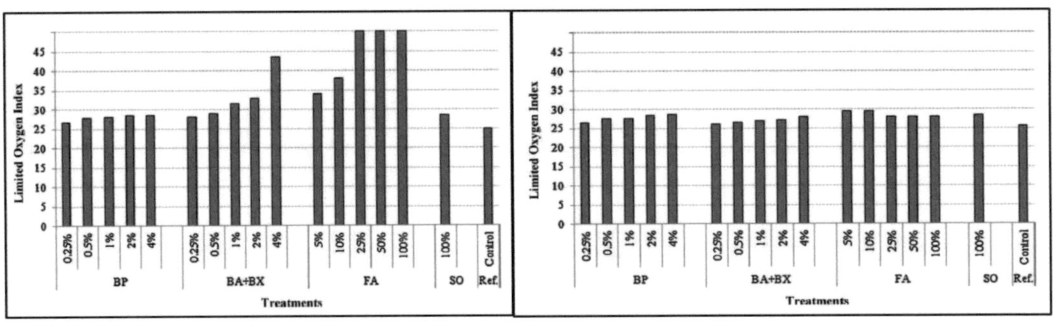

Figure 3. OI of un-leached and leached flame retardant-treated wood [38].

OI of wood. However, OI of treated sample was affected by leaching procedure, which may be due to the preservatives are not chemically adhered to the wood [38].

Disodium octaborate tetrahydrate ($Na_2B_8O_{13}.4H_2O$) in waterborne paint treatment provides better fire retardant properties of Scot pine wood. The properties are further improved by tannin-based wood preservative solutions [39]. Further, the effect of wood preservative loading on the OI of fir wood was investigated. Wood preservatives have been widely used to extend the service life of wood. Study indicated that almost all treated samples showed higher OI values, and moreover, 3% copper-based preservative (Wolmanit-CB) was recommended [40]. The improvement in dimensional stability, durability and fire retardant properties of wood was investigated by using 1,3-dimethylol-4,5-dihydroxyethyleneurea treatment. The Oak wood showed considerable decrease in the pyrolysis temperature, heat release rate, and smoke production when methylolguanylurea phosphate and boric acid flame retardants were incorporated [41]. There is a current contentious toxicity problem with boron compounds, and as such, a need for wood products to move away from boron-based fire retardants is recognized.

8. Phosphorus-based fire retardant coatings

Fire retardant treatment on wood with flame retardants of both nitrogen compounds and phosphoric acid was examined. The fire retardance and endurance of wood were influenced by method of treatments, such as heat pressed treatment and heat dried treatment. The heat pressed treatment method was improved the properties of wood as compared to heat dried treatment. It was found that the flame retardant properties were further improved by the amount and functional reactivity of flame retardants with formaldehyde as in the dicyandiamide-formaldehyde-phosphoric acid or melamine-dicyandiamide-formaldehyde-phosphoric acid system. The endothermic release of water during the condensation of phosphoric acid can also cool the wood and dilute the volatile pyrolysis gases [42]. Similarly, Subyakto et al. also demonstrated the improvement in fire retardant properties of wood by phosphoric acid treatment, preheating and densifying the surface of wood. Trimethylol melamine formaldehyde resin mixed with phosphoric acid was coated on the wood surface, which was preheated and followed by hot pressing. The pressurized impregnation of coating was improved the fire retardancy of wood without reduction in the bending strength [43]. The fire retardancy of white pinewood was improved by treatment with orthophosphoric acid at different concentration. The aqueous solution of thiourea-formaldehyde resin and orthophosphoric acid was impregnated at different concentration for 1 h. It was found that the weight, compressive strength and fire retardant properties were improved after impregnation. However, the water uptake of treated wood was increased in a water soaking test for 168 h [44].

Organophosphorus flame retardant compounds are not often used in commercial wood applications, and some are known to have high volatility. The synergism between phosphorus and nitrogen is also observed in organophosphorus compounds. Rupper et al. investigated the surface chemistry of cellulose treated with fire retardants containing ethyl ester phosphoramidates. Evidence for a condensed-phase action of the fire retardants was found [45]. It was also showed that phosphoramides, when bonded to cellulose, increase the char yield and lower the weight loss rates in comparison to phosphorus pentoxide and amine-treated wood [46]. The modified pine sawdust with alkyl and phenyl chlorophosphorus compounds using pyridine was prepared by Stevens et al. They reported a reduction in the temperature of maximum pyrolysis rate of up to 90°C and an increase in char formation of up to 29%. The efficiency of the phenyl phosphates was favored compared to the alkyl analogues, and the order of effectiveness was, expectedly, attributed to the acidity and thermal stability, that is, phosphate > phosphonate > phosphinate [47]. Further, it was showed that coupled with a copper-based preservative, the impregnation of wood with an organophosphorus fire retardant reduced the FIGRA by a further 15% [48].

In another study, wood was treated with guanidine compounds, such as guanidine dihydrogen phosphate, diguanidine hydrogen phosphate, guanidine carbonate, and guanidine nitrate and analyzed thermal degradation properties by using thermogravimetric analysis. Char yields were increased compared to untreated wood by approximately 60, 55, 20 and 25%, respectively, which effectively demonstrates the synergism by phosphorus and nitrogen based fire retardants [49]. The effect of urea-formaldehyde oligomer reacted nitrogen-phosphorus flame retardant treatment on dimensional stability and the flame retardant properties of wood was investigated. The results showed that both dimensional stability and OI values were improved significantly in flame retardant impregnated wood. The better flame retardant properties of treated wood are due to formation of protective char layer through dehydration of polysaccharides, which terminate both heat and oxygen [50].

Stejskal et al. studied the flame retardant properties of wood coated with polyaniline, and the coating was made in hydrochloric or phosphoric acid solutions in the absence and presence of stabilizers, poly(n-vinylpyrolidone) or colloidal silica. The coated wood showed less mass loss and formation of charcoal layer on the surface when exposed to direct flame or in a furnace temperature at 400–600°C, as compared to uncoated wood. The similar observation was made with polypyrole and poly(1,4-phenylenediamine) as deposition polymers in wood. The soaking of wood in polyaniline colloids was badly affected the flame retardant properties, whereas the reaction between the cellulose fibers and polyaniline was required to enhance the stability of wood at high temperature. This is attributed to the formation of carbonaceous microtubes, which offered the higher stability of wood against flame and heat exposure [51]. Cyclophosphazene is a polymeric material containing both nitrogen and phosphorus, which has wide range of thermal and chemical stability in addition with fire retardant properties. El-Wahab et al. synthesized the three kinds of cyclodiphosh(V)azane compounds (I-III) and physically mixed in polyurethane varnish formulation at different concentration. It was found that OI of wood panels was increased with loadings (**Figure 4**). They claimed that improvement was mainly due to several factors, the high molecular weight and aromatic cyclophosphazenes containing chlorine, nitrogen and phosphorus that provides superior flame retardant properties. The presence of N-P bonds renders exceptionally thermally stable, and release less toxic and corrosive gases during burning [52].

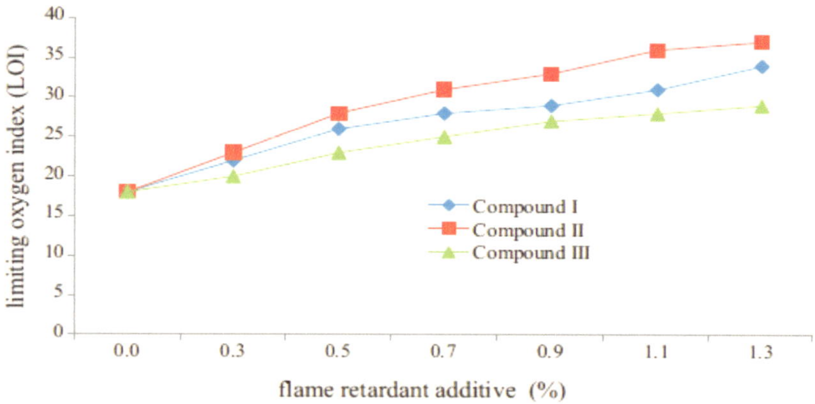

Figure 4. OI of wood coated with varnish containing cyclodiphosh(V)azane compounds [52].

9. Coatings with synergistic flame retardants

The combinations of silicon and phosphorus have proved popular in the fire retardancy community, in addition to silicon, phosphorus and nitrogen mixtures. The synergism is generally explained as a combination of the individual effects of each of the three additives: phosphorus provides the effective char formation, nitrogen produces non-combustible gases acting as diluents, and silicon offers thermal stability to the substrate by forming a protective layer over the forming char throughout decomposition. The white deposit of silicon dioxide covering the surface of the char will act as a radiant heat shield and help to reduce the rate of oxidation of the char [53]. Grexa and Lubke studied the effect of magnesium hydroxide (MDH), monoammonium phosphate (MAP), aluminum hydroxide (ATH), and boric acid on the flammability of particle board using cone calorimeter at external irradiance of 50 kW/m^2. The combination of MDH, MAP and boric acid showed better fire retardant properties, in terms of both heat release rate and smoke production as compared to the same composition contains ATH instead of boric acid. It is expected that while the phosphorus and nitrogen will have behaved synergistically to direct the pathway of pyrolysis toward more char and water and fewer flammable volatiles, the boron present will have become molten to form a glass-like barrier on the surface of the wood, stabilizing the char and enforcing the mass transport barrier. Further, the expandable graphite-based intumescent flame retardants showed lower heat release rate and mass loss rate of particle board when compared to the same loading of ammonium polyphosphate. It was found that the char layer forming flame retardants have a strong effect of flame retardation of wood. The forming char has a distinctive effect in the performance of the material when exposed to external heat in comparison to the same flame retardant without a char layer. They suggested that the intumescent flame retardant system has potential to be used to improve the reaction-to-fire performance of wood [54, 55].

Canosa et al. studied the role of reinforcing fibers on the flammability of intumescent flame retardant coated wood panel. Different film forming materials were chosen to be blended with active ingredients, pigment and several fibers, alumina, carbon, aramid, and glass fibers.

Results suggested that chlorinated rubber, phenolic, and epoxy resin showed best performance in thermal conductivity test, 2-foot tunnel, OI and UL94 horizontal-vertical test and also fibers achieved the synergistic effect with intumescent coatings [56]. A 5A-zeolite-treated ammonium polyphosphate (APP) showed better fire performance properties, heat release rate (HRR), total heat released (THR), smoke production rate (SPR), total smoke released (TSR) and fire growth index (FGI) of sawdust board (SB) as compared to APP being used alone (**Table 1**). Further improvement of fire performance was observed by acid (4-picolinic acid) impregnated 5A-zeolite-treated APP. This can be because acid significantly altered the thermal decomposition and catalytically decomposed to retard the combustion, which promotes the uniform char formation [57].

Carbonates and hydrogen carbonates are known to have very high efficiency as gas-phase flame retardants. Potassium carbonate is reported as a compound with a high fire retardant efficiency for wood products. It has a relatively high decomposition temperature (800°C) and serves as a catalyst to the dehydration of wood to increase the production of char, water and CO_2. However, the compound is unable to prevent the depolymerization of wood effectively, particularly at high concentrations, and also causes the evolution of CO. As such, the potassium carbonate is only used in low concentrations [58, 59]. The effects of various inorganic salts, Na_2WO_4, Na_2SnO_3, Na_2MoO_4 on the thermal decomposition and fire retardant properties of wood sample were demonstrated. Upon the treatment of these salts, the OI of wood sample was increased, which is caused by an increase in the amount of char on the surface. The activation energies of the samples were also decreased after treatment during both the charring stage and the calcining stage. The flame retardants were shown to be able to catalyze the dehydration reaction, resulting in the formation of more H_2O, CO_2, and char, but less flammable vapors like levoglucosan and levoglucose [60].

The effects of pressure and microwave heating duration on the flammability of ammonium polyphosphate (APP) impregnated wood samples were studied. The flame retardant properties, such as peak heat release rate, total heat released and total smoke released, were measured for samples of pretreated and untreated with microwave and characterized by cone calorimeter. It was found that the treated wood showed better flame retardant properties, and moreover, the microwave pretreatment of wood can also increase the fire retardant properties of APP impregnated wood [61]. Wang et al. found that the expanded vermiculite treatments improve the flame retardant properties of plywood. Results showed that expanded vermiculite treatments increase the OI values of wood and at the same time decrease the thermal activation energy at the maximum degradation process. The increase in OI is due to the formation of protective layer on the surface of wood sample [62].

Sample	Peak HRR (kW/m²)	Time to peak HRR (s)	Average HRR (kW/m²)	THR (MJ/m²)	SPR (m²/s)	TSR (m²/m²)	FGI (kW/sm²)
SB-APP	122.0	40	29.2	4.4	310	4.1	5.3
SB-APP/5A	124.8	60	24.5	3.3	229	2.0	3.1
SB-APP/5A	50.1	50	8.5	1.1	215	1.9	1.4

Table 1. Cone calorimetric date for sawdust board [57].

10. Intumescent fire retardant coatings

The intumescent coating of either pigmented or clear was introduced to the fire retardation of wood-based products. Intumescent coatings swell and char when exposed to heat, giving carbonaceous foam that insulates the surface from the fire. The char layer is also responsible for the limitation of oxygen diffusion and the reduction of the volatilization of the fuel in order to prevent the continuation of the combustion cycle. Intumescence resulted from the application of flame retardant coating systems to timber can reduce the char formation and heat buildup and also delay the onset of combustion of a wood [63]. Gardner and Thomson have studied the flammability of forest products, including sawn boards, plywood, hardboards and particleboard as per ASTM E906–Standard test method for heat and visible smoke release rates for materials and products. The intumescent flame retardant was used to treat the plywood by pressure impregnation. The ignition time of treated sawn board was not increased with density when exposed to heat fluxes of 20 and 40 kW/m^2. However, the ignition times of plywood were increased by pressure impregnated flame retardants. The heat release properties of both samples were reduced by flame retardant coatings, which is dependent on the exposed heat flux [64].

An intumescent fire-retardant coating based on unsaturated polyester and epoxy resin was prepared by using ammonium polyphosphate, pentaerythritol, melamine, and expandable graphite, and studied the fire endurance performance of wooden board. Results showed that 2 mm film thickness of intumescent coating provides excellent fire endurance time [65]. Chou et al. investigated the usefulness of an artificial graphite powder and a sericrite ($Al_4(OH)_4(KAl\text{-}Si_3O_{10})_2$), and a mixture of the two on plywood. The intumescent fire retardant coating formulation contains 19.8% of the flame retardant, 15% of the dehydration agent, 18% of the foaming agent, 7.2% of the resin binder and 40% of the solvent which was prepared and applied to the surface of plywood. They showed that when sericrite was in excess of 75% in the fire retardant composition, the mixture obtained the lowest flammability grade possible in Taiwan Standard CNS 7614. Furthermore, for sericrite to be effective in inhibiting combustion and also carbonizing agents, such as graphite powder, are not required [66]. The burning behavior of intumescent coating-treated flax board was studied by cone calorimeter and single burning item (SBI) test. Intumescent coating provides significant increase in ignition time and decrease in heat release rate of flax board. It was also confirmed that experimental results are comparable to numerical predictions, factional factor, the ratio of the heat flux at the interface of the intumescent surface and the char layer of flax board to the surface heat flux when the absence of intumescent coating layer, based on analytical solutions for charring materials and burning rates in SBI tests [67]. Intumescent coatings are commonly used for protecting steel structures. Their application on wood has also been studied, and some commercial products intended for wood are available. At present, however, fire protection of wooden and wood-based products with intumescent coatings is not widely employed.

11. Transparent fire retardant coatings

The transparent ultraviolet (UV) curable intumescent fire retardant coating on wood was prepared by using cyclotriphosphazene as a flame retardant. A series of (2-hydroxyethylmethacrylate) (n-propoxy)cyclotriphosphazenes were prepared by the reaction of $N_3P_3Cl_6$ with n-propanol and 2-hydroxyethylmethacrylate sequentially in the appropriate solvents. It was observed that phosphazene-based compounds showed better fire retardant properties of wood and without affecting the aesthetic appearance of wood structure [68]. Chen et al. prepared the dual cured (UV-radiation and moisture) flame retardant coatings based on silicone and phosphate modified acrylates. The oligomer was made by reacting 9,10-dihydro-9-oxa-10-phosphaphenanthrene-10-oxide and 3-glycidoxypropyltriemethoxysilane, and reacted with 2,4-toluene diisocyanate and 2-hydroxyethyl acrylate. The dual cured film showed high-flame retardance, which is attributed to the synergistic effect of phosphorus-silicon and phosphorus-nitrogen [69]. Shi and Wang synthesized the transparent intumescent flame retardant coatings from epoxy/phosphate flame retardant and amino resins. Five types of flame retardants were prepared by a two-step reaction using 1-oxo-4-hydroxymethyl-2,6,7-trioxa-1-phosphabicyclo[2.2.2]octane, polyphosphoric acid, bisphenol-A epoxy resin and 1,3-butanediol diglycidyl ether with different proportions. The structure of flame retardant was confirmed by nuclear magnetic resonance (^{1}H-NMR) and Fourier transforms infrared (FTIR) spectroscopy. It was found that the fire retardant properties of plywood were significantly improved by intumescence and also enable to maintain the visibility of the surface features of wood [70].

12. Nanocomposite coatings

Nanoparticles have recently been used to prepare the nanocomposites for the improvement in fire retardant properties. The major concern of these materials is dispersion. The surface modification is essential for nanoparticles to achieve better compatible and homogeneous dispersion. The necessary loading of nanoparticles is usually lower than for their micron-sized filler counterparts which are an enormous advantage industrially and economically. Clay nanopowder composed of montmorillonite and cellulose nanofibers which is used as a transparent fire retardant coating for wood was demonstrated. Fire performance was assessed by cone calorimeter at irradiance of 35 kW/m^2. It was found that nanopowder coated wood showed a strong increase in time to ignition, and a 33% total heat release was reduced along with the 46% reduction in maximum average heat emission rate. Both thermal shielding and gas barrier functions contribute to delayed thermal degradation of wood and delayed emission of volatile combustible gases [71]. Hassan et al. prepared the flame retardant intumescent polyurethane coatings for wood products. The limitation of intumescent additives, such as incompatibility and loading issue, was addressed by using butyl acrylate and montmorillonite clay. They found that the flame retardant properties of wood were improved by addition of acrylate and nanoclay using cone calorimeter [72]. Giudice and Pereyra showed through the use of oxygen index and a two-foot tunnel tests, (ASTM E84 2013—the regulatory test for interior flammability of building materials in the United States), the silica nanoparticle-treated wood provides several significant advantages: high fire retardancy, low thermal expansion, reduced smoke production, and low cost [73]. The reaction-to-fire properties of titanium dioxide (TiO_2) and/or clay nanoparticles coated spruce wood were investigated by Fufa et al. using a small-scale cone calorimeter. They found that the negative results of reaction-to-fire performance along with water vapor permeability were observed on specimens treated with TiO_2 and/or clay nanoparticles treatments [74]. Chuang et al. have investigated the fire performance of intumescent coated plywood with the addition of commercial organoclays (Cloisite 30B, Cloisite 10A, and Cloisite 15A) at different loadings, 1, 3, 5, and 10%. According to the cone calorimeter test, compared to uncoated plywood, the intumescent coating exhibited lower peak heat release rate (peak HRR) and extend the time to reach peak HRR. Further improvement was observed with the incorporation of organoclays; however, the type and amount of organoclay are very important. For instance, 3% Cloisite 30B and 5% Cloisite 10A displayed better fire retardancy as compared to other loadings of the clay. It was demonstrated that adding organoclay extends the survival duration of the phosphocarbonaceous char structures, which is confirmed by spectroscopic studies [75].

Zhang et al. studied the ignition and burning behavior of intumescent coating and sepiolite nanoparticles applied on flaxboard using cone calorimeter, single burning item (SBI) test, and reduced scale (one-third) ISO room test. Both cone calorimeter and SBI test represent an open and well-ventilated condition, whereas reduced ISO room test is a fire in a confined space and represents a more realistic burning condition in most compartment fires. Intumescent coatings effectively delaying the ignition and mass loss rate in both cone calorimeter and SBI tests. Further improvement was observed in addition of nanoparticles. This because of intumescent coating forms carbonaceous char on the surface, which acts as a thermal and physical barrier preventing heat, mass and gas transfer, and further improvement is due to an increase in the thermal stability of the char as confirmed by the thermogravimteric analysis. In addition, it was also confirmed that intumescent coatings do not cause an increase of the toxic gases, and ventilation plays a vital role in the development of the fire [76].

The high thermal conductivity of nanosilver coatings was tested to improve heat transfer in wood and enhance fire retardant performance. Nanosilver treatment clearly showed potential in improving some of the fire retarding properties in solid wood products. It can be observed that coating may delay thermal degradation and carbonization by reducing the accumulation of heat that is rapidly transferred [77]. Nano-wollastonite is demonstrated as multifunctional additive in wood. Wollastonite nanofibers were also reported the same group to improve durability and fire retardant properties of poplar wood and solid wood composites. It was found that both fire retardant and dimensional stability were improved at 10% nano-wollastonite impregnated wood. As a mineral material of wollastonite, acts as an impermeable physical barrier toward the penetration of flames into the wood structure [78]. The synthesis of hexagonal boron nitride nanosheets was demonstrated through a facile shear force liquid phase exfoliation method and used as a binder free oxidation and fire retardant wood coatings. Because of intrinsic low thermal diffusivity and thermal effusivity, nanosheet coatings showed an excellent fire retardation and oxidation resistance up to 900°C [79]. The effect of organically modified alpha-zirconium phosphate (OZrP) on the thermal and fire retardant properties UV curable system was studied. The flame retardant coating system comprises of phenyl di(acryloyloxyethyl)phosphate (PDHA), triglycidylisocyanurate acrylate (TGICA) and 2-phenoxyethyl acrylate (PHEA). Results showed that addition of 0.5 wt% OZrP, the peak heat release rate and total heat of combustion were reduced significantly. This is because of the effective char formation and improvement of anti-oxidation performance of the coating [80].

13. Sol-gel method

In recent years, sol-gel processes have also become recognized for the purposes of incorporating fire retardants into products. The process comprises hydrolysis and condensation reactions that lead to the formation of inorganic or organic-inorganic hybrid coatings. This technique is well documented for different polymers. Giudice et al. synthesized the polysiloxanes in wood pores by sol-gel process using aminopropylmethyldiethoxysilane, aminopropyltriethoxysilane and a mixture of both (50/50 ratio), and then, impregnated panels were subjected to 2-foot tunnel test (flame spread index, panel consumption, and smoke density). Impregnation process was carried out at 40–50°C in an autoclave and controlling the operating conditions for achieving different weight gains. It was shown by the authors that aminopropyltriethoxysilane-treated wood sample showed best fire retardant efficiency. This is because of more reactivity of alkoxide, which forms hybrid structure [81]. Another study showed that the transparent fire retardant coating for wood (pine and larch) was prepared by a sol-gel method using vinyl functionalized zirconium oxy-clusters copolymerized with vinyl trimethoxysilane. Results showed that coating has improved the fire retardant properties and also without affecting the macroscopic appearance of wood surface [82].

14. Fire resistance of timber

The use of timber components in the loading structure in a building relies on fire engineering design to ensure that the building can retain its structural integrity for sufficient time either for building occupants to be evacuated, or for the fire to be extinguished. In construction using large cross-section timber members, like cross-laminated timber, this may be done by assuming a rate at which the timber chars and therefore the cross-section of timber remaining after a given time [83]. The strength and stiffness of timber both reduce at lower temperatures than steel and concrete. For example, timber's strength is reduced by more than 50% at 100°C, compared with that at 20°C [84]. Timber structural members may still perform well at high temperatures in comparison with steel, however, since the char layer can act to insulate the material within, whereas the high thermal conductivity of steel means that the complete section quickly heats up. Where steel is used to connect timber elements, heat can be quickly conducted through the connectors, degrading the strength and stiffness of the wood around them. The behavior of timber in fire is fundamentally different to steel and reinforced concrete; however, since it is combustible, research groups have identified that the key research needs to be addressed for the next generation of large timber buildings [85, 86]. They address the performance of systems with various levels of encapsulation, the effect of flame spread due to a combustible structural material, and the fire performance of connections. Another potential use for coatings is to increase the fire resistance of structural timbers. It was showed that significant increases in fire resistance can be achieved by using fire resistant coatings. This concept should be pursued by the timber industry, as it would be possible to improve the fire resistance of structural timbers in old buildings that are being remodeled. The application of a fire resistive coating would be simpler and cheaper than cladding the members in fire resistive board materials, or replacing the timber member with a concrete or steel member.

http://dx.doi.org/10.5772/67675

Author details

Thirumal Mariappan

Address all correspondence to: thirumal.mariappan@asianpaints.com

Research and Technology Center, Asian Paints Limited, Navi Mumbai, Maharashtra, India

References

[1] Weil ED. Fire-protective and flame-retardant coatings—a state-of-the-art review. Journal of Fire Sciences. 2011;**29**:259-295

[2] Bourbigot S, Duquesne S. Fire retardant polymers: recent developments and opportunities. Journal of Materials and Chemistry. 2007;**17**:2283-2300

[3] Mariappan T. Recent developments of intumescent fire protection coatings for structural steel: A review. Journal of Fire Sciences. 2016;**34**:120-163

[4] White RH, Dietenberger MA. Fire Safety of Wood Construction. Wood Handbook. Madison, WI: U.S. Department of Agriculture, Forest Service;1999. p. 463

[5] Jimenez M, Duquesne S, Bourbigot S. Multiscale experimental approach for developing high-performance intumescent coatings. Industrial & Engineering Chemistry Research. 2006;**45**:4500-4508

[6] Verburg GB, Rayner ET, Yeadon DA, et al. Water-resistant, oil-based, intumescing fire-retardant coatings. I. Developmental formulations. Journal of the American Oil Chemists' Society. 1964;**41**:670-674

[7] Wilkie CA, Morgan AB. Fire Retardancy of Polymeric Materials, 2nd ed. New York: CRC Press; 2010

[8] Östman BA.-L, Mikkola E. European classes for the reaction to fire performance of wood-based panels. Fire and Materials. 2010;**34**:315-330

[9] Goldsmith FP. Fire retardant coatings : an evaluation of fire retardant coatings as a means of protecting wood panels. doi:http://dx.doi.org/10.14288/1.0103122, Wood 493, April 17, 2011

[10] Hirata T, Kawamoto S, Nishimoto T. Thermogravimetry of wood treated with water-insoluble retardants and a proposal for development of fire-retardant wood materials. Fire and Materials. 1991;**15**:27-36

[11] Pabeliña KG, Lumban CO, Ramos HJ. Plasma impregnation of wood with fire retardants. Nuclear Instruments and Methods in Physics Research Section B. 2012;**272**:365-369

[12] Janssens M. Piloted ignition of wood: A review. Fire and Materials. 1991;**15**:151-167

[13] Zicherman JB, Allard DL. Fire performance of fire-retardant wood fiberboard ceiling tile. Fire and Materials. 1992;**16**:187-196

[14] Buchanan AH. Fire performance of timber construction. Progress in Structural Engineering and Materials. 2000;**2**:278-289

[15] Östman BA.-L, Voss A, Hughes A, Hovde PJ, Grexa O. Durability of fire retardant treated wood products at humid and exterior conditions review of literature. Fire and Materials. 2001;**25**:95-104

[16] Lowden LA, Hull TR. Flammability behaviour of wood and a review of the methods for its reduction. Fire Science Reviews. 2013;**2**:1-19

[17] Barber D, Gerard R. Summary of the fire protection foundation report - fire safety challenges of tall wood buildings. Fire Science Reviews. 2015;**4**:1-15

[18] Marzi T. Nanostructured materials for protection and reinforcement of timber structures: A review and future challenges. Construction and Building Materials. 2015;**97**:119-130

[19] Maraveas C, Miamis K, Mathew CE. Performance of Timber Connections Exposed to Fire: A Review. Fire Technology. 2015;**51**:1401-1432

[20] Shi L, Chew MYL. A review of fire processes modeling of combustible materials under external heat flux. Fuel. 2013;**106**:30-50

[21] Leung CW, Chow WK. REVIEW ON FOUR STANDARD TESTS ON FLAME SPREADING. International Journal on Engineering Performance-Based Fire Codes. 2001;**3**:67-86

[22] Ramage MH, Burridge H, et al. The wood from the trees: The use of timber in construction. Renewable and Sustainable Energy Reviews. 2017;**68**:333-359

[23] Fudang S, Zhiming DU, et al. Experimental Study on Fires Extinguishing Properties of Melamine Phosphate Powders. Procedia Engineering. 2014;**84**:535-542

[24] Li Q, Jiang P, We P. Synthesis, characteristic, and application of new flame retardant containing phosphorus, nitrogen, and silicon. Polymer Engineering and Science. 2006;**46**:344-350

[25] Shi L, Chew MYL. Experimental study of carbon monoxide for woods under spontaneous ignition condition. Fuel. 2012;**102**:709-715

[26] Xue-qiu Z, Guo-qing Z, Guo-xiang Z. Experimental Study on Wood Material under Different Oxygen Concentration. Procedia Engineering. 2013;**52**:607-612

[27] McAllister S. Critical mass flux for flaming ignition of wet wood. Fire Safety Journal. 2013;**61**:200-206

[28] Maciulaitis R, Lipinskas D, Lukosius K. Singularity and Importance of Determination of Wood Charring Rate in Fire Investigation. Materials Science. 2006;**12**:42-47

[29] Beikircher W, Hartmann P, Kogl J. World Conference on Timber Engineering; Canada. 2014

[30] Atar M, Keskin H. Impacts of coating with various varnishes after impregnation with boron compounds on the combustion properties of Uludag fir. Journal of Applied Polymer Science. 2007;**106**:4018-4023

[31] Keskin H. Effects of impregnation materials on combustion properties of laminated veneer lumber obtained from European oak (Quercus petraea Liebl.) and lombardy poplar (Populus nigra L.). Journal of Applied Polymer Science. 2007;**105**:1766-1773

[32] Blasi CD, Branca C, Galgano A. Flame retarding of wood by impregnation with boric acid – Pyrolysis products and char oxidation rates. Polymer Degradation and Stability. 2007;**92**:752-764

[33] Branca C, Blasi CD. Oxidation characteristics of chars generated from wood impregnated with (NH4)2HPO4 and (NH4)2SO4. Thermochimica Acta. 2007;**456**:120-127

[34] Blasi CD, Branca C, Galgano A. Effects of Diammonium Phosphate on the Yields and Composition of Products from Wood Pyrolysis. Industrial & Engineering Chemistry Research. 2007;**46**:430-438

[35] Blasi CD, Branca C, Galgano A. Thermal and catalytic decomposition of wood impregnated with sulfur- and phosphorus-containing ammonium salts. Polymer Degradation and Stability. 2008;**93**:335-346

[36] Branca C, Blasi CD. Semi-global mechanisms for the oxidation of diammonium phosphate impregnated wood. Journal of Analytical and Applied Pyrolysis. 2011;**91**:97-104

[37] Baysal E, Altinok M, Colak M, et al. Fire resistance of Douglas fir (Pseudotsuga menzieesi) treated with borates and natural extractives. Bioresource Technology. 2007;**98**:1101-1105

[38] Tomark ED, Cavdar AD. Limited oxygen index levels of impregnated Scots pine wood. Thermochimica Acta. 2013;**573**:181-185

[39] Tondi G, Haurie L, Wieland S, et al. Comparison of disodium octaborate tetrahydrate-based and tannin-boron-based formulations as fire retardant for wood structures. Fire and Materials. 2014;**38**:381-390

[40] Cavdar AD. Effect of various wood preservatives on limiting oxygen index levels of fir wood. Measurement. 2014;**50**:279-284

[41] Jiang T, Feng X, Wang Q, et al. Fire performance of oak wood modified with N-methylol resin and methylolated guanylurea phosphate/boric acid-based fire retardant. Construction and Building Materials. 2014;**72**:1-6

[42] Getto H, Ishihara S. Functionally graded wood in fire endurance with basic nitrogen compounds and phosphoric acid. Fire and Materials. 1998;**22**:77-83

[43] Subyakto, Kajimoto T, Hata T, Ishihara S, et al. Improving fire retardancy of fast growing wood by coating with fire retardant and surface densification. Fire and Materials. 1998;**22**:207-212

[44] Mahmoud AA, Eissa AMF, Omar MS, et al. Improvement of white pinewood properties by impregnation with thiourea–formaldehyde resin and orthophosphoric acid. Journal of Applied Polymer Science. 2000;**77**:390-397

[45] Rupper P, Gaan S, Salimova V, Heuberger M. Characterization of chars obtained from cellulose treated with phosphoramidate flame retardants. Journal of Analytical and Applied Pyrolysis. 2010;**87**:93-98

[46] Lee HL, Chen GC, Rowell RM. Thermal properties of wood reacted with a phosphorus pentoxide–amine system. Journal of Applied Polymer Science. 2004;**91**:2465-2481

[47] Stevens R, Van Es DS, Bezemer R, Kranenbarg A. The structure–activity relationship of fire retardant phosphorus compounds in wood. Polymer Degradation and Stability. 2006;**91**:832-841

[48] Marney DCO, Russell LJ. Combined Fire Retardant and Wood Preservative Treatments for Outdoor Wood Applications – A Review of the Literature. Fire Technology. 2008;**44**:1-14

[49] Gao M, Ling B, Yang S, Zhao M. Flame retardance of wood treated with guanidine compounds characterized by thermal degradation behavior. Journal of Analytical and Applied Pyrolysis. 2005;**73**:151-156

[50] Jiang J, Li J, Gao Q. Effect of flame retardant treatment on dimensional stability and thermal degradation of wood. Construction and Building Materials. 2015;**75**:74-81

[51] Stejskal J, Trchová M, Brodinová J, Sapurina I. Flame retardancy afforded by polyaniline deposited on wood. Journal of Applied Polymer Science. 2007;**103**:24-30

[52] El-Wahab HA, El-Fattah MA, El-Khalik NA, et al. Synthesis and performance of flame retardant additives based on cyclodiphosph(V)azane of sulfaguanidine,1,3-di-[N/-2-pyrimidinylsulfanilamide]-2, 2, 2.4, 4, 4-hexachlorocyclodiphosph(V)azane and 1,3-di-[N/-2-pyrimidinylsulfanilamide]-2, 4-di[aminoacetic acid]-2, 4-dichlorocyclodiphosph(V)azane incorporated into polyurethane varnish. Progress in Organic Coatings. 2012;**74**:615-621

[53] Kandola BK, Horrocks AR, Price D, Coleman GV. Flame-Retardant Treatments of Cellulose and Their Influence on the Mechanism of Cellulose Pyrolysis. Journal of Macromolecular Science. 1996;**36**:721-794

[54] Grexa O, Lubke H. Flammability parameters of wood tested on a cone calorimeter. Polymer Degradation and Stability. 2001;**74**:427-432

[55] Grexa O, Poutch F, Manikova D, et al. Intumescence in fire retardancy of lignocellulosic panels. Polymer Degradation and Stability. 2003;**82**:373-377

[56] Canosa G, Alfieri PV, Giudice CA. Hybrid Intumescent Coatings for Wood Protection against Fire Action. Industrial & Engineering Chemistry Research. 2011;**50**:11897-11905

[57] Yuan L, Chen X, Hu Y. Combination effect of 4-picolinic acid with 5A zeolite on ammonium polyphosphate flame-retarded sawdust board. Journal of Fire Sciences. 2014;**32**:230-240

[58] Dobele G, Urbanovich I, Zhurins A, et al. Application of analytical pyrolysis for wood fire protection control. Journal of Analytical and Applied Pyrolysis. 2007;**79**:47-51

[59] Kadir O, Abdullah CI, Erol B, Salih A. The effect of potassium carbonate, borax and wolmanit on the burning characteristics of oriented strandboard (OSB). Construction and Building Materials. 2007;**21**:1457-1462

[60] Qu H, Wu W, Wu H, Jiao Y, Xu J. Thermal degradation and fire performance of wood treated with various inorganic salts. Fire and Materials. 2011;**35**:569-576

[61] He X, Li XJ, Zhong Z, et al. Effectiveness of impregnation of ammonium polyphosphate fireretardant in poplar wood using microwave heating. Fire and Materials. 2016;**40**:818-825

[62] Wang F, Gao Z, Zheng M, Sun J. Thermal degradation and fire performance of plywood treated withexpanded vermiculite. Fire and Materials. 2016;**40**:427-433

[63] Richardson LR, Cornelissen AA. Fire-resistant coatings for roof/ceiling deck timbers. Fire and Materials. 1987;**11**:191-194

[64] Gardner WD, Thomson CR. Ignitability and heat-release properties of forest products. Fire and Materials. 1991;**15**:3-9

[65] Gu J, Zhang G, Dong S, et al. Study on preparation and fire-retardant mechanism analysis of intumescent flame-retardant coatings. Surface and Coatings Technology. 2007;**201**:7835-7841

[66] Chou CS, Lin SH, Wang CI. Preparation and characterization of the intumescent fire retardant coating with a new flame retardant. Advanced Powder Technology. 2009;**20**:169-176

[67] Zhang J, Delichatsios MA, McKee M, Ukleja S. Experimental and numerical study of burning behaviors of flaxboard with intumescent coating and nanoparticles in the cone calorimeter and single burning item tests. Fire and Materials. 2012;**36**:554-564

[68] Chen-Yang YW, Chuang JR, Yang YC, et al. New UV-curable cyclotriphosphazenes as fire-retardant coating materials for wood. Journal of Applied Polymer Science. 1998;**69**:115-122

[69] Chen X, Hu Y, Jiao C, Song L. Preparation and thermal properties of a novel flame-retardant coating. Polymer Degradation and Stability. 2007;**92**:1141-1150

[70] Shi Y, Wang G. The novel epoxy/PEPA phosphate flame retardants: Synthesis, characterization and application in transparent intumescent fire resistant coatings. Progress in Organic Coatings. 2016;**97**:1-9

[71] Carosio F, Cuttica F, Medina L, Berglund LA. Clay nanopaper as multifunctional brick and mortar fire protection coating—Wood case study. Materials & Design. 2016;**93**:357-363

[72] Hassan MA, Kozlowski R, Obidzinski B. New fire-protective intumescent coatings for wood. Journal of Applied Polymer Science. 2008;**110**:83-90

[73] Giudice CA, Pereyra AM. Silica nanoparticles in high silica/alkali molar ratio solutions as fire-retardant impregnants for woods. Fire and Materials. 2010;**34**:177-187

[74] Fufa SM, Steen-Hansen A, Jelle BP, et al. Reaction to fire and water vapour resistance performance of treated wood specimens containing TiO2 and clay nanoparticles. Fire and Materials. 2014;**38**:717-724

[75] Chuang C-S, Tsai K-C, Yang T-H, et al. Effects of adding organo-clays for acrylic-based intumescent coating on fire-retardancy of painted thin plywood. Applied Clay Science. 2011;**53**:709-715

[76] Zhang J, Delichatsios M, McKee M, et al. Experimental study of burning behaviors of intumescent coatings and nanoparticles applied on flaxboard. Journal of Fire Sciences. 2011;**29**:519-530

[77] Taghiyari HR. Fire-retarding properties of nano-silver in solid woods. Wood Science and Technology. 2012;**46**:939-952

[78] Poshtiri AH, Taghiyari HR, Karimi AN. The optimum level of nano-wollastonite consumption asfire-retardant in poplar wood (Populus nigra). International Journal of Nano Dimension. 2013;**4**:141-151

[79] Liu J, Kutty RG, Zheng Q, et al. Hexagonal Boron Nitride Nanosheets as High-Performance Binder-Free Fire-Resistant Wood Coatings. Small. 2017;**13**:1602456

[80] Xing W, Zhang P, Song L, et al. Effects of alpha-zirconium phosphate on thermal degradation and flame retardancy of transparent intumescent fire protective coating. Materials Research Bulletin. 2014;**49**:1-6

[81] Giudice CA, Alfieri PV, Canosa G. Siloxanes synthesized "in situ" by sol–gel process for fire control in wood of Araucaria angustifolia. Fire Safety Journal. 2013;**61**:348-354

[82] Girardi F, Cappelletto E, Sandak J, et al. Hybrid organic–inorganic materials as coatings for protecting wood. Progress in Organic Coatings. 2014;**77**:449-457

[83] Matthew W. Tall Timber Buildings: Applications of Solid Timber Construction in Multistory Buildings. CTBUH J. 2011;**1**:24-27

[84] BS EN 1995-1-2:2004. Eurocode 5. Design of timber structures. Part 1-2: General. Structural Fire Design, 15 December 2004.

[85] Andrew B, Östman B, Andrea F. Fire Resistance of Timber Structures. Technical Report, National Institute of Standards and Technology; March 2014. Gaithersburg, Maryland, United States.

[86] Robert G, David B, Armin W. Fire Safety Challenges of Tall Wood Buildings. Technical Report, The Fire Protection Research Foundation, QUINCY, MASSACHUSETTS, U.S.A. December 2013.